BUSINESS ECONOMICS IN A RAPIDLY-CHANGING WORLD

THE ROLE OF CORPORATE VENTURE CAPITAL IN INNOVATION

Business Economics in a Rapidly-Changing World

The Role of Risk and Information in Consumption and Savings
Antonella Tutino
2009. ISBN: 978-1-60741-263-2

Joint Ventures, Mergers and Acquisitions, and Capital Flow
James B. Tobin and Lawrence R. Parker (Editors)
2009. ISBN: 978-1-60692-795-3

Strategic Economics of Network Industries
Hans Gottinger
2009. ISBN: 978-1-60741-345-5

Backdating of Stock Options
Bernard T. Weinstein (Editor)
2009. ISBN: 978-1-60741-984-6

Mark to Market Accounting Standards: A Study by the SEC
Brian N. Brinker (Editor)
2009. ISBN: 978-1-60741-994-5

Mark to Market and Fair Value Accounting: An Examination
James W. Curtis (Editor)
2009. ISBN: 978-1-60741-996-9

International Accounting in the 21st Century
Aleksei Solovyov and Ilia Kuznetsov (Editor)
2009. ISBN: 978-1-60876-018-3

The Role of Corporate Venture Capital in Innovation
Joseph F. Tollington (Editors)
2010. ISBN: 978-1-60876-936-0
2010. ISBN: 978-1-61668-427-3 (E-book)

Multiple Criteria Decision Aiding
Constantin Zopounidis, Michael Doumpos, Nikolaos F. Matsatsinis and Evangelos Grigoroudis (Editors)
2010. ISBN: 978-1-61668-231-6
2010. ISBN: 978-1-61668-606-2 (E-book)

BUSINESS ECONOMICS IN A RAPIDLY-CHANGING WORLD

THE ROLE OF CORPORATE VENTURE CAPITAL IN INNOVATION

JOSEPH F. TOLLINGTON
EDITOR

Nova Science Publishers, Inc.
New York

For permission to use material from this book please contact us:
Telephone 631-231-7269; Fax 631-231-8175
Web Site: http://www.novapublishers.com

LIBRARY OF CONGRESS CATALOGING-IN-PUBLICATION DATA

The role of corporate venture capital in innovation / editor, Joseph F. Tollington.
p. cm.
Includes index.
ISBN 978-1-60876-936-0 (hardcover)
1. Venture capital. 2. Technological innovations--Finance. 3. Corporations--Finance. 4. New business enterprises--Finance. I. Tollington, Joseph F.
HG4751.R65 2010
338'.064--dc22

2010012293

Published by Nova Science Publishers, Inc. † New York

CONTENTS

PREFACE

Venture capital (also known as VC or Venture) is a type of private equity capital typically provided for early-stage, high-potential, growth companies in the interest of generating a return through an eventual realization event such as an IPO or trade sale of the company. Venture capital investments are generally made as cash in exchange for shares in the invested company. It is typical for venture capital investors to identify and back companies in high technology industries such as biotechnology and ICT (information and communication technology). This new book details the role of venture capital in innovation.

Chapter 1 - This study examines the relationship between corporate investments and the degree of internationalization by companies receiving venture capital funds. In 2000, more than $100 billion in venture capital disbursements, nearly one-fifth of which came from corporations, helped ease some of the resource constraints that affect 98 percent of entrepreneurs in technology-based ventures. That same year, nearly 200,000 small technology ventures conducted international sales. Prior research disagrees about whether the receipt of venture capital is related to increased levels of international activity. However, these studies have not explicitly considered the use of corporate venture capital, which provides differential value-added from that of independent venture capital. This study of 268 new technology-based ventures finds that companies that receive corporate venture capital (CVC)—that is, equity investments by established corporations— are larger, older, better funded, and further along in their development than ventures that have not received corporate funding. Controlling for these differences, a positive and significant relationship exists between the receipt of CVC and a higher percentage of revenues earned from foreign sources. However, there is no conclusive evidence that either corporate international diversity or prior international investing experience is a mechanism by which this relationship exists. This study suggests that entrepreneurs following internationalizing strategies consider corporate sources of funding and that independent venture capital providers consider corporate syndicate partners for their portfolio companies interested in foreign market entry.

This study represents one portion of my doctoral dissertation. Another portion examines whether venture capital providers manage their investments differently if the venture undertakes foreign activities. A third paper studies the relationship between internationalization and venture performance at initial public offerings.

Chapter 2 - This report examines corporate venture capital (CVC) as a model of innovation. CVC programs in established corporations invest in and partner with entrepreneurial companies. By doing so, established companies are able to identify and source new emerging technologies from entrepreneurial companies.

CVCs typically make a financial investment and receive a minority equity stake in an entrepreneurial company. CVCs also facilitate investment of in-kind resources into portfolio companies. In return, the parent corporation gains a window on new technologies and strategically complementary companies that could become strategic partners. CVCs generally invest with a combination of financial and strategic objectives. Strategic objectives include leveraging external sources of innovation, bringing new ideas and technologies into the company, and taking "real options" on technologies and business models (by investing in a wider array of technologies or business directions than the company can pursue itself).

Corporate venture capital may be viewed in the broader context of corporate venturing, including both internal and external venturing. Internal venturing programs "go inside" the firm and create entrepreneurial ventures from within the corporation. External venturing programs "go outside" the firm and tap external sources of innovation, whether through research collaborations with universities, strategic alliances with other firms, or partnerships with entrepreneurial companies. Often, the firm's internal and external venturing efforts are closely related and interact with each other.

CVC programs in established corporations face both inward and outward. They face outward to build relationships with the entrepreneurial venture community, learn about new technology and business directions, and make investments that create new strategic opportunities for the corporation. They face inward to interact with the firm's R&D and business operating units, in order to identify operating units' interests and priorities. CVCs support the corporation's existing businesses by introducing new technologies and partnerships to its operating groups. At the same time, CVCs help identify technologies and opportunities that fall between or beyond the corporation's existing businesses.

This report uses industry data and original survey data to describe trends and characteristics of CVC organizations and investments. These data provide insight on a range of issues relating to CVC operations and investments.

In: The Role of Corporate Venture Capital in ... ISBN: 978-1-60876-936-0
Editors: Joseph F. Tollington

Chapter 1

CORPORATE VENTURE CAPITAL AND THE INTERNATIONAL INTENSITY OF PORTFOLIO COMPANIES

Joseph A. LiPuma
SBA Office of Advocacy

EXECUTIVE SUMMARY

This study examines the relationship between corporate investments and the degree of internationalization by companies receiving venture capital funds. In 2000, more than $100 billion in venture capital disbursements, nearly one-fifth of which came from corporations, helped ease some of the resource constraints that affect 98 percent of entrepreneurs in technology-based ventures. That same year, nearly 200,000 small technology ventures conducted international sales. Prior research disagrees about whether the receipt of venture capital is related to increased levels of international activity. However, these studies have not explicitly considered the use of corporate venture capital, which provides differential value-added from that of independent venture capital. This study of 268 new technology-based ventures finds that companies that receive corporate venture capital (CVC)—that is, equity investments by established corporations— are larger, older, better funded, and further along in their development than ventures that have not received corporate funding. Controlling for these differences, a positive and significant relationship exists between the receipt of CVC and a higher percentage of revenues earned from foreign sources. However, there is no conclusive evidence that either corporate international diversity or prior international investing experience is a mechanism by which this relationship exists. This study suggests that entrepreneurs following internationalizing strategies consider corporate sources of funding and that independent venture capital providers consider corporate syndicate partners for their portfolio companies interested in foreign market entry.

This study represents one portion of my doctoral dissertation. Another portion examines whether venture capital providers manage their investments differently if the venture undertakes foreign activities. A third paper studies the relationship between internationalization and venture performance at initial public offerings.

INTRODUCTION

This study examines the relationship between corporate investments and the degree of internationalization by ventures receiving venture capital funds. Resource constraints affect 98 percent of entrepreneurs (Evans & Jovanovic, 1989). More than \$100 billion in venture capital disbursed in 2000, nearly one-fifth of which came from corporations, helped to ease those constraints. That same year, more than 210,000 firms exported; the vast majority were small technology ventures, earning over \$150 billion in export revenue (International Trade Administration, 2003). Maula and Murray (2002) found that entrepreneurs perceive that corporate investors provide more value to new venture internationalization strategies. I find a positive relationship between the receipt of corporate equity funding and a higher degree of international intensity—defined here as the percentage of revenues a company obtains from foreign sales— compared to ventures receiving equity investment solely from noncorporate sources.

Young international ventures, in contrast to those following solely domestic strategies, utilize different types of resource stocks (Oviatt & McDougall, 1994). Internationalizing firms incrementally acquire one such resource, experiential knowledge, as they conduct business in foreign environments (Kogut & Zander, 1993). Other sources of this knowledge are the prior experience of the top management team (Reuber & Fisher, 1997), board of directors (Bloodgood, Sapienza & Almeida, 1996), and venture capital providers (Carpenter, Pollock & Leary, 2003).

Venture capital (VC) provides new companies with the knowledge, experience, and network resources of the funding organization. Such VC resources may influence the geographic goals of new ventures (Gupta & Sapienza, 1992), many of which choose to internationalize. International new ventures (Oviatt & McDougall, 1994) are similar in age and industry profile to customary recipients of VC investments. The intensity with which they internationalize is "considerable," with revenues from foreign sales often constituting more than 25 percent of their total revenues (Knight & Cavusgil, 2005).

The knowledge, networks, and experience of VC providers affect the value-added associated with venture capital. Independent and corporate investors provide differential value: the former nurtures internal processes, whereas the latter supports external relationships (Maula & Murray, 2002). In 2000, 335 U.S. corporations provided nearly $20 billion in venture capital, a twentyfold increase since 1997 (Rauser, 2002). Studies indicate that new venture performance is positively associated with the receipt of corporate VC (Gompers & Lerner, 1998). The increase in the number of corporate venture capital (CVC) investors, the amount of corporate investment, and the significant international activities by small firms[1] raise the question: "what is the relationship between the receipt of CVC and the international intensity of new ventures?"

Prior studies of experience and internationalization have focused on the top management team (Reuber & Fisher, 1997) or alliance partners (Sharma & Blomstermo, 2003); VC providers may take on either or both roles with the portfolio companies in which they invest. Few rigorous empirical studies of the value-added by VC focus on the perspective of the portfolio company receiving the funds, although notable exceptions include Gompers and Lerner (1998) and Maula and Murray (2001). Little research exists regarding the relationship between VC and the intensity of internationalization.

Studies of the relationship between the receipt of VC and the degree to which ventures internationalize have either been inconclusive (e.g., Burgel & Murray, 1998), or have found a negative relationship (e.g., Carpenter et al., 2003). However, these studies have not considered the source of the capital, despite indications that CVC provides more value-added in an international context (Maula & Murray, 2001). The effect of this differential value-added on international intensity may be masked in studies that do not control for source, potentially leading to contradictory findings. By including a variable for corporate investment in the analysis, I am able to empirically test whether VC sources are related to international intensity.

VC providers and the companies in which they invest are generally spatially concentrated to enhance relationship formation and facilitate monitoring (Sorenson & Stuart, 2001), increasing the potential for success of the venture (Green, 2004). Most CVC investors are multinational enterprises (MNEs) such as Intel and Microsoft (Rauser, 2002).[2] Their global reach may increase the radius of exchange with their portfolio companies (PCs), and their ability to build relationships and monitor their investments may exceed that of independent venture capital (IVC) providers, who have only recently begun to internationalize (Manigart, Collewaert, Wright, et al., 2006).

To address the question of the relationship between the receipt of CVC and the international intensity of new ventures, I identified a sample of 268 young, privately owned technology companies that received VC between 1997 and 2003, roughly half of which had international sales in that period. Although only 4 percent of U.S. companies export,[3] the high percentage (48 percent) of ventures with foreign sales in this sample is consistent with prior research indicating a higher likelihood of international activities for technology firms (Jones, 1999). I analyzed the relationship between their international intensity in 2003 and their sources of equity investments by using regression models that address potential selection bias. Heckman (1979) two-stage models correctly deal with the fact that not all ventures in this study selected an international strategy. Neither of the previously cited studies of the relationship between the receipt of VC and the degree of internationalization (i.e., Burgel & Murray, 1998; Carpenter et al., 2003) address selection bias, despite the inclusion of solely domestic ventures in their samples. Explanatory variables included prior international investing experience, venture age at first investment, number of CVC investments and CVC parent international presence. The findings indicate that the chances of a CVC-backed venture deriving more than 10 percent of its revenue from foreign sources are almost 60 percent greater than the chances for a venture not similarly funded.

This finding is important for academics and practitioners. Scholars in both entrepreneurship and international business disciplines are trying to reconcile the existence of international new ventures (INVs) with extant theory. Studies of new ventures often use the intensity of international activities as a condition for considering such ventures INVs. Identifying determinants of international intensity enhances analytic rigor that informs theory and aids subsequent empirical investigations. For entrepreneurs with options when selecting investors, the finding of a relationship between corporate investors and high international intensity may provide insights for use in the selection process. Similarly, investors may (re)consider the inclusion of corporate funds in their investment syndicates based on how they view the value of venture internationalization.

The organization of this paper is as follows. The next section provides a theoretical development of the value of corporate venture capital and its ability to confer a competitive advantage resulting in increased international intensity. The development of hypotheses regarding the nature of such a relationship precedes a description of analytic methods for testing these hypotheses. Presentation of test results paves the way for a discussion of implications and limitations of the study that concludes the paper.

THEORY AND HYPOTHESES

International business theories suggest that firms internationalize when they possess a monopolistic advantage (Caves, 1996), in the presence of other ownership, location, or internalization advantages (Dunning, 1988), or for control (Hymer, 1976; Kogut & Zander, 1993). International new ventures—businesses that, from inception, seek to derive significant competitive advantage from the use of resources and the sale of outputs in multiple countries (Oviatt & McDougall, 1994:49)—often possess neither the advantages nor the resources for control. However, many such firms exist—Knight and Cavusgil (2005) identified more than 1,000 "born global" ventures in the United States that derive more than 25 percent of their revenues from foreign sources. Prior research has found a positive relationship between international intensity and the prior international experience of the top management team (TMT) and board of directors (Bloodgood et al., 1997; Reuber & Fisher, 1997). Research has produced mixed findings regarding the relationship between international intensity and receipt of VC (e.g., Burgel & Murray, 1998; Carpenter et al., 2003) despite numerous studies of the knowledge and network value-added of VC providers. Corporate investors, in particular, possess resources that have the potential to provide significant value-added to internationalizing ventures (Maula & Murray, 2001).

The rapidity and intensity with which firms internationalize are central to the definition of INVs for studies of this phenomenon. A consensus is developing that INVs are six or fewer years old and derive more than 10 percent of their revenues from foreign sources (Coviello & Jones, 2004). Identifying the age at which ventures initiate cross-border activities and the factors causing such activities, and identifying resources that might influence the intensity of these efforts may facilitate development of a typology of such resources, helping to explain the existence of INVs.

International intensity is apparent at all firm size levels; companies with high degrees of internationalization are both large and small (Calof, 1993). While early foreign market entry is associated with faster international growth (Autio, Sapienza & Almeida, 2000), Preece, Miles & Baetz (1998) conclude that it is the availability of resources necessary to pursue international sales that influences international intensity. Although they considered only financial resources, resources such as market learning (Yeoh, 2004) can also influence international intensity.

In young firms, the human capital (e.g., knowledge and experience) of TMT represents a significant corporate asset. Start-ups are more likely to have international activities if their founders had prior international education or work experience (Burgel & Murray, 1998; McDougall, Oviatt & Shrader, 2003). Such experience of the TMT facilitates internalization of market knowledge (Yeoh, 2004) and relates to the degree of internationalization via the timing of foreign market entry and the process by which it is accomplished (Reuber & Fisher, 1997).

The prior international experience of members of the board of directors increases awareness of opportunities in foreign environments, and positively relates to a firm's extent of internationalization (Bloodgood et al., 1997). This association is strongest when both the TMT and outside directors (including venture capitalists) have prior international experience (Carpenter et al., 2003). Such experience may serve as a proxy for the reduction of uncertainty and as a surrogate for accumulating cultural knowledge, and this experience may be inimitable and nonsubstitutable (Daily, Certo & Dalton, 2000).

Venture capital provides knowledge, advice, and active managerial support in addition to financing (Gorman & Sahlman, 1989; Hellmann & Puri, 2002). Venture capital provides access to human capital and social capital via the knowledge, networks, and other valuable capabilities of the funding organization. Such resources can help PCs as they move beyond their national borders, as specialized market knowledge is crucial to market entry, and international networks can substitute for resource constraints in these situations (Coviello & McAuley, 1999). Sapienza (1992) concludes that the role of strategy and advice taken by VC providers is facilitated by frequent communications with venture founders/managers, which enhances knowledge transfer and internalization.

The value-added activities of VC providers are knowledge-based (e.g., product development assistance, business and marketing plan formulation) and network-based (e.g., investor relations, customer and distributor contact) (Smith, 2001). CVC providers help in both regards, deriving greater technical knowledge and market insights from their embeddedness in MNEs. This embeddedness helps to deal with the added complexity of the international environment, resulting in greater value-added to internationalization. Further, while all internationalizing firms experience liabilities of foreignness in nondomestic settings (Zaheer, 1995), associating with a prominent MNE partner (Stuart, Hoang & Hybels, 1999) can offset the additional liability of newness experienced by young ventures (Stinchcombe, 1965). Such legitimacy positively influences market acceptance and sales of the foreign venture.

Information and knowledge are perhaps the most critical resources to the international expansion of the firm (Knight & Liesch, 2002); tacit knowledge in particular is difficult to obtain and is a strategic asset in the achievement of company goals (Winter, 1987). Tacit knowledge, that which is complex and difficult to codify or teach, includes foreign market and cultural knowledge (Kogut & Zander, 1993). Acquisition of such tacit knowledge occurs over time, and firm structures and routines facilitate its internal transference.

In "traditional" international firms, relevant resources for internationalization tend to be highly tangible (e.g., plant, property, equipment) or are largely composed of financial and human resources, whereas INVs tend to utilize intangible resources, primarily special knowledge about specific strategies and approaches for international business (Knight & Cavusgil, 2005).

Venture capital organizations may also employ their prior international investing experience to the benefit of internationalizing PCs. The involvement of VC providers with the TMT of companies can directly (via board of director positions) or indirectly (via advice and networks) influence the international intensity of these new ventures. The ability of VC providers to supply advice and assistance with decision-making is knowledge-based (Grant, 1996; Kogut & Zander, 1992). Knowledge is the strategically most significant asset of the firm (Grant, 1996), and internalization of knowledge of international environments is the basis for internationalization (Kogut & Zander, 1993). The primary source of a firm's international knowledge lies within the prior international experiences of its management team (Grant, 1996), its investors and advisors.

Maula and Murray (2001) conclude that while IVC providers' value is "enterprise nurturing" (e.g., strategy development, financing assistance, and executive recruiting), CVC providers help with "commerce building" (e.g., increasing public credibility, helping to attract customers, suppliers, and partners). CVC providers, via their parent MNEs, have international networks, adding to the number of weak network ties of their PCs. The possible speed for increasing international scope and the number of cross-border weak ties that an entrepreneurial actor has established are related (Oviatt & McDougall, 2005); such scope can result in greater intensity of international activities.

The current study seeks to fill a gap in the literature by determining whether a particular type of relationship, that between an international new venture and a corporate investor, is related to a high degree of international intensity. Such a finding would contribute to the understanding of why INVs exist.

Hypotheses

This study investigates whether a correlation exists between different characteristics of venture financing and the degree of internationalization: in the case of venture capital, how does CVC, the prior experience of the CVC provider, and the existence of an international marketing and operations capability relate to high levels of PC international intensity. I apply the resource-based view (Barney, 1991) as the theoretical lens for this analysis. This approach is consistent with prior suggestions as to its suitability for studies of INVs (Oviatt & McDougall, 1997).

The resource-based view examines utilization of resources that are rare, valuable, inimitable, and nonsubstitutable. Venture capital is rare because it requires mutual opportunity recognition by entrepreneurs and VC providers to enact investments. In the United States in 2000, the most active VC investment year in history, approximately 6,700 companies (roughly 0.1 percent of firms founded in the previous decade[4]) received such funding (Green, 2004). Approximately 1 percent of business plans submitted result in obtaining VC financing.[5] CVC is even more rare; fewer than 13 percent of deals in 2000 included corporate investment.[6]

Since VC providers generally do not invest in multiple firms in the same product/service niche, any other firm receiving VC will have an imperfect imitation of the nonfinancial (e.g., knowledge, experience, network) resources provided by the VC source. Such resources have a strong tacit dimension and social complexity (Alvarez & Busenitz, 2001), such as the unique relationship established between the entrepreneur and the VC provider, rendering them inimitable. In essence, the entrepreneur purchases a right to the scarce resources offered by the VC provider.

Knowledge based on prior international experiences results in a rare and inimitable resource for a company (because of the unique historical context of each experience) that enables it to overcome the liability of foreignness (Zaheer, 1995). The uncertain imitability resulting from the difficulty of identifying performance causality (Lippman & Rumelt, 1982) makes it complicated to find a substitute for such international experience.

Different sources of investment capital result in different investment objectives, compensation schemes, capabilities and risks, each of which has implications for the companies receiving funds. IVC providers maintain strictly financial objectives, whereas most CVC organizations also invest for strategic reasons, seeking to obtain indirect benefits in addition to, or instead of financial gains (Gompers & Lerner, 2000; Rauser, 2002). Strategic goals may take the form of development of complementary technology that may benefit the MNE parent in international markets. With CVC, those providing the investment capital are in the same corporation with those making investment decisions and monitoring those investments; this fact may reduce agency problems and improve communication (Maula & Murray, 2001).

More than IVC firms, CVC organizations are valuable in helping PCs obtain new foreign customers (Maula, Autio & Murray, 2005). Corporations typically have specialist knowledge in industry sectors related to their own. CVC organizations also have access to nonmaterial technical, market, and business knowledge that IVC firms rarely possess (Maula & Murray, 2001), enhancing the opportunity recognition of their PCs. By providing young portfolio companies with market and technical knowledge (Kann, 2001), CVC providers can enhance the knowledge in young firms, due to their "learning advantage of newness" (Autio et al., 2000), facilitating the growth of the PC's business in foreign markets, and increasing their international intensity.

A firm is an institution for integrating knowledge (Grant, 1996). The institutional environment of the CVC embedded in a firm with related knowledge provides more trust (Uzzi, 2000) and helps the PC to learn more quickly about international markets. This learning assistance helps portfolio companies gain incremental knowledge of foreign environments, allowing them to make successively greater commitments and potentially move more rapidly through stages of internationalization (Johanson & Vahlne, 1977) or circumvent them altogether.

Frequent communication between CVC investors and their PCs, and their hands-on approach to providing value (McNally, 1997) facilitate learning. The value of such knowledge- based resources depends upon the ability to value that knowledge through experience, and to assimilate and apply it (Cohen &

Levinthal, 1990). Such abilities are facilitated by the venture's absorptive capacity based, in part, on an overlap in technical knowledge between the VC provider and PC (Oxley & Sampson, 2003), which is greater with CVC than with IVC.

A startup's relationship with its corporate investor can enhance its ability to form additional alliances (Kelley & Spinelli, 2001). Networks of the VC provider represent a valuable resource to the portfolio company, providing value in the pursuit of foreign market opportunities (Coviello & Munro, 1995) and in the pursuit of other required resources (Maula et al., 2005).

Venture capital adds value to new ventures and CVC adds greater value than IVC relative to helping ventures attain a high degree of internationalization. Therefore, new technology-based ventures that receive CVC funding are more likely to have high international intensity.

Hypothesis 1: Among VC-backed new ventures that are international, those receiving corporate venture capital are more likely than those not receiving such CVC to have a higher level of international intensity.

Characteristics of both investors and ventures can intensity or dampen the effect that the receipt of corporate VC has on international intensity. Not all corporations are equal with respect to their ability to contribute to substantial internationalization of ventures in which they invest. Ventures that receive CVC from firms possessing certain characteristics are more likely to have a high degree of internationalization. Ventures are similarly heterogeneous with respect to their ability to utilize investor contributions. Ventures possessing certain characteristics are more likely to attain high international intensity with CVC funding.

Investor characteristics and new venture internationalization

Companies that operate in an international network may enjoy a "learning advantage" and find it easier to go abroad than companies that operate only in domestic networks (Sharma & Blomstermo, 2003). Young firms, in particular, have a "learning advantage of newness" and may be better able to learn and adapt to new foreign environments (Autio et al., 2000), facilitating faster international growth. Thus, ventures whose investors have previously funded and presumably maintained ties with internationalized ventures may have such an advantage.

Entrepreneurs wishing to internationalize can benefit from the human capital of a VC provider with international experience (Manigart et al., 2006).

Opportunity recognition may be greater for ventures backed by such VC organizations because of their prior knowledge (Shane, 2000). Prior experience with international investments provides a corporate investor with knowledge of how to communicate with and monitor such ventures, even at a distance from the ventures' activities.

The prior international experience of VC providers is a resource that confers an advantage in internationalizing. Venture capital organizations are generally small, averaging fewer than 30 general partners. Organizations of this size are more apt to share information and have an awareness of the backgrounds and activities of their staff; thus the international experience of the VC organization as a whole will provide value to the new venture. Inimitability of the prior experience stems from the unique historical and geographic context in which the activity occurred and a high degree of tacit knowledge inherent in the prior experience.

Information and knowledge are critical to international expansion (Knight & Liesch, 2002), and there are limited substitutes for the knowledge that prior experience of an investor brings to a relationship. Investors have a strong interest in using their human and social capital to the benefit of the portfolio companies in which they invest, and the application of such capital reduces search costs for opportunities, facilitating sales.

Not all corporations are equal in their experience with foreign ventures. Ventures that receive CVC from firms with prior experience investing in foreign ventures are more likely to have a high degree of internationalization.

> *Hypothesis 2: Conditional on receiving CVC funds, new ventures financed by CVC providers with greater prior international investing experience will be more likely to have a high level of international intensity.*

The lack of technical and marketing experience are two major obstacles faced by internationalizing companies (Brush, 1992). Most CVC units are in major industrial corporations (Rauser, 2002), mainly multinational enterprises.[7] Such multinationals maintain international subsidiaries and have an extensive network of partners worldwide. More than one-half of all U.S. MNEs operate in five or more countries (International Trade Administration, 2003).

Providing managerial or technical assistance to entrepreneurial ventures is characteristic of approximately 80 percent of CVC programs (Kann, 2001). Such assistance is based on a linkage to the investing company's operational capabilities (i.e., its resources and processes). PCs might make use of that company's manufacturing plants, distribution channels, technology, pricing benefits or brand, or may adopt the investing company's business practices (Rauser, 2002). Such capabilities, valuable in their own right, also lead to development of larger networks for the new venture. VC providers display lower preferences for geographic diversification because of the difficulty in providing assistance long distance; transnationality compounds this difficulty (Gupta & Sapienza, 1992). However, such networks expand the radius of exchange (Sorenson & Stuart, 2001) and act as proxies for venture capital localization, which facilitates venture success (Green, 2004).

Large MNEs also conduct market research that may be valuable for new ventures operating in related fields (Maula & Murray, 2001). The greater the degree to which these additional value- adding capabilities are in other countries, the greater the potential advantage for INV PCs.

Not all corporations possess a diverse international presence to contribute to substantial internationalization of ventures in which they invest. Ventures that receive CVC from firms with a high degree of international diversity are more likely to have a high degree of internationalization.

> *Hypothesis 3: Conditional on receiving CVC funds, new ventures financed by corporations with a diverse foreign presence will be more likely to have a high level of international intensity.*

Venture characteristics and new venture internationalization

New firms have a "learning advantage of newness" (Autio et al., 2000), because they lack entrenched routines. They may more easily internalize early knowledge of foreign markets afforded by VC providers, resulting in the development of routines supportive of foreign market penetration. New ventures can glean technology and knowledge of market needs from CVC providers who, via their parent MNEs, have extensive foreign networks of suppliers, partners, and customers. Early knowledge of these needs, when the venture is developing its offerings, results in products and services that more readily conform to foreign needs, facilitating sales in foreign environments.

Most CVC is provided in later rounds, when the venture is older and in late stages of development (at which times products have been tested and the venture is focused more on expanding its presence). Although rare, CVC in early rounds, when the company is younger, results in increased assistance by the CVC parent in intensifying international sales of the venture.

Ventures are heterogeneous with respect to their ability to utilize investor contributions. Ventures at a younger age with fewer embedded routines and a "learning advantage of newness" when they first receive CVC funds are more likely to have a high degree of internationalization.

Hypothesis 4a: Conditional on receiving CVC funds, new ventures receiving venture funding at a younger age are more likely to have a high level of international intensity than those receiving such funding at a later age.

Figure 1 illustrates the relationships proposed in the preceding hypotheses.

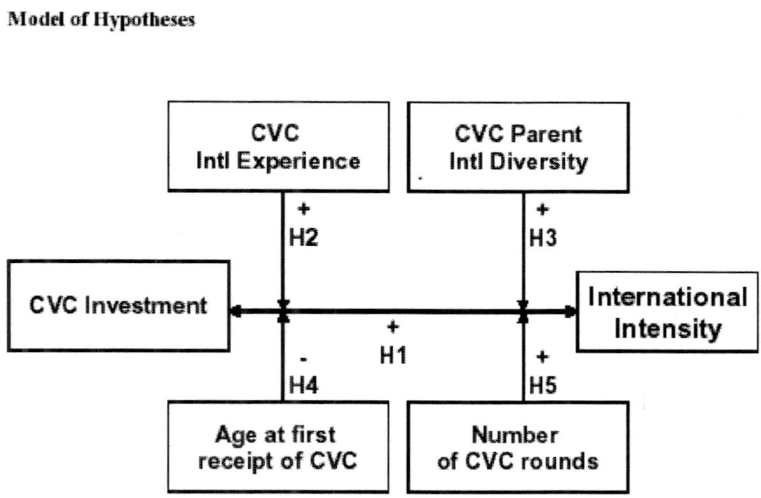

Figure 1. Model of Hypotheses

While communications regularly take place between investors and ventures, the number of face-to-face meetings relates to perceived VC value (Sapienza, 1992). Venture capital providers use funding rounds as a means of both managing their risk and providing knowledge and advice to their PCs. Such information sharing can include both explicit and tacit knowledge. Explicit knowledge may include identification of opportunities, names of client prospects, and information on potential alliance partners. Not every member of the TMT can obtain tacit knowledge directly; groups develop a common understanding via face-to-face interactions (Athanassiou & Nigh, 2000). Contact and interaction in financing rounds facilitates the transfer of tacit market knowledge to the TMT of new ventures. More rounds of financing provide more contact and communication, and thus provide opportunity for information exchange of this type. Such information can aid the new venture in identifying and developing foreign opportunities and increasing foreign sales.

Not all ventures that obtain CVC receive the same number of rounds that include corporate funding. Ventures receiving more rounds of CVC are more likely to have a high degree of internationalization.

> *Hypothesis 5: Conditional on receiving CVC funds, new ventures receiving more rounds of corporate funding are more likely to have a high level of international intensity than those receiving fewer such funding rounds.*

METHODS AND ANALYSIS

Only a subset of the sample exhibits the dependent variable, *High Intensity*, as not all companies in the sample have international sales, suggesting that the criteria for selecting observations are not independent of the outcome variable. I thus used a Heckman selection procedure (Heckman, 1979) to address the possibility of selection bias in the sample. This procedure first estimates the likelihood of a firm conducting international activities based on a set of characteristics (selection model) and then evaluates how specific factors relate to a high degree of international intensity (effects model), controlling for factors that may also relate to high intensity. Entrepreneurs will self-select into the geographic strategy (domestic/international) that provides a better match with their abilities and hence a greater return. Without modeling this self-selection, a regression of high intensity on funding source may lead to erroneous estimates for intensity. Significant *F* values and *Rho* statistics confirmed that the chosen two-stage Heckman regressions were appropriate.[8]

The first stage of the Heckman procedure, the selection model, uses a probit regression to estimate the likelihood of having had international sales for the full sample (n=268). The second stage of the Heckman procedure, the effects model, incorporates the estimates into a probit regression model to predict the probability of high international intensity of ventures that had internationalized by 2003 (n=129).

Sample

To test the hypotheses, I extracted data from CorpTech, which maintains data on U.S.-based companies that manufacture, do research and development, or provide services relating to high technology products. CorpTech data has previously been used for sample selection (e.g., Harveston, Kedia & Davis, 2000; Knight & Cavusgil, 2004), identification of company size (Simonin, 1999), and verification of sample representativeness (Soh, 2003). An advantage of CorpTech is that it seeks to include data from all firms in more than 700 technology-based 4-digit SIC codes. A broad range of industries, albeit primarily technology-based, is thus included. CorpTech updates data annually utilizing surveys, phone interviews, and analyses of publicly available data from SEC filings, industry reports, press releases, and other media. CorpTech may estimate some data elements, such as detailed sales figures and number of employees, based on available information. I excluded from my sample companies that declined to provide relevant data.

Beginning with a random sample of 7,506 companies obtained from CorpTech in September 2005, I identified those ventures that were privately held (5,679). I eliminated those companies that were not primarily high-technology-based, resulting in 5,539 companies. Within this set, I identified ventures for which an historical path was available in the CorpTech data for the years 2002, 2000, and 1998.[9] This process resulted in an identification of 1,368 companies, 824 of which had founding dates between 1997 and 2003 (inclusive). A key research design question is how to choose firms that are of the appropriate age. I focus on ventures six or fewer years old because firm survival is determined in this period (*The State of Small Business*, 1992) and VCs disburse over 80 percent of VC rounds to similarly aged companies.[10] Selection of ventures of this age is also consistent with previous literature on INVs (Brush, 1995; Coviello & Jones, 2004).

I matched the names of the 824 companies meeting the selection criteria against a listing of all companies receiving VC between 1997 and 2003 as identified by VentureXpert/SDC Platinum by Thompson Financials (henceforth SDC). This database has an extensive research history relative to CVC (e.g., Gompers & Lerner, 1999) and is one of the most comprehensive sets of data on VC activity.

I eliminated six companies from the resultant set of 370 because of activities prior to 1997 or the nature of VC transactions (e.g., acquisitions) in the period 1997-2003. I eliminated an additional six companies due to missing VC disbursement data. The resultant 358 companies received 1,218 rounds of

VC, representing 4,489 investor deals. More than one-third (129) of these companies had a known, non-0 percentage of their revenues derived from foreign sources, and 139 were purely domestic. These 268 ventures represent the sample for this study.

The sample contains only survivors—a common issue in studies involving data spanning multiple years. Given the bursting of the Internet bubble, it is conceivable that many technology ventures founded in the 1990s were not around in 2003. On the other end of the spectrum, the sample also eliminates the most successful firms, i.e., those founded in 1997 or later with an IPO prior to 2004. Thus, the study looks at the group of companies that did not die, but did not achieve a successful IPO exit for investors. The effect this has on the study is dependent on how the market views internationalization at IPO. If investors view intense internationalization positively, ventures with high international intensity will have undergone an IPO, biasing the sample toward low- intensity ventures. The "survivor" bias in the sample might lean toward high intensity, as early internationalization leads to increased mortality, but also increased growth for survivors (Sapienza, Autio, George & Zahra, 2006).

Data

CorpTech's dataset includes the following for each company: founding date, percentage of foreign sales (five categories: 0, <2.5%, 2.5-10%, 10-25% and 25+ %), sales revenue, industry, and number of employees. Because CorpTech obtains data from different sources on different dates, it provides a separate "as of" date for both the sales figures and the number of employees. In all cases, these dates were within one year of each other; in half the cases, these dates were within four months of each other.

I obtained financing histories for the matched companies from SDC, including company data (name, founding date and stage), round data (number, date, disclosed size, number of investors and total number received) and investor data (name, nation, and type (e.g., corporation). For the sample period 1997-2003 (inclusive), SDC reports data on 5953 firms that provided more than $688 million in VC investments. Independent VC providers represented more than 50 percent of the investors (3,106 firms), followed by 728 corporate subsidiaries or affiliates.

I used the Directory of Corporate Affiliations to obtain data on the number of countries in which the CVC parent operated foreign operations (subsidiaries, affiliates and joint ventures) in 2001. This data source has an extensive history in research for determining the extent of internationalization of multinationals (e.g., Carpenter & Sanders, 2004; Chen & Martin, 2001).

Measures[11]

The dependent variable, *High Intensity,* reflects whether the percentage of revenue a company obtains from foreign sales exceeds 10 percent. The selection model censors out those ventures that had no international activity, permitting an analysis of ventures with high versus low international intensity. *High Intensity* takes a value of "1" if the focal venture earned more than 10 percent of its revenue from foreign sources in 2003 or "0" if this ratio of foreign sales to total sales (FSTS) is nonzero but less than 10 percent. My use of FSTS is consistent with prior research (e.g., McDougall & Oviatt, 2003; Preece et al., 1998), as FSTS addresses the overall value of international business to the venture via the depth of its involvement in, importance of, and commitment to foreign operations.

The key independent variable of interest in this study is *CVC*. CVC, corporate venture capital, is a dummy variable that takes a value of "1" if the focal venture received CVC from a U.S. firm any time in its history and "0" otherwise. I also include a number of independent variables that measure relevant CVC characteristics: C *VC Experience*, *CVC Presence*, *Age 1st CVC,* and *CVC Rounds*. C *VC Experience* is a measure of the experience that the VC organization develops by prior investing in foreign ventures, and is a count of the number of rounds of investment that CVC providers to the focal company invested in foreign firms during the period 1994-2002.

The variable *CVC Presence* represents a count of the number of foreign countries in which CVC investors in the focal firm had subsidiaries in 2001. In the event that more than one CVC investor provided funds to the focal company, I use the larger of the number of unique countries in which the multiple firms operate. Corporations require time to obtain market knowledge and establish market presence used for the benefit of ventures in which they invest. A two-year lag between the measure for this variable and that of the dependent variable is consistent with previous research suggesting lags for factors that may affect performance (e.g., George, Wiklund & Zahra, 2005). *Age 1st VC* is the age, in months, at which the focal company first received venture capital. Receipt of CVC at an early stage may initiate foreign market information internalization at a time when the venture has few embedded routines. The variable *CVC Rounds* is a count of the number of rounds of venture capital received by the focal firm in which a U.S. corporation participated.

To address other factors that might relate to high international intensity, it is important to include a number of control variables. International business process theories (e.g., Johanson & Vahlne, 1977) suggest that the degree of internationalization is a function of age of the firm; thus, a control for venture *Age* is included. *Age* is the number of years between the venture founding date and the reported date of sales.

Considering the resource demands of internationalization, the total amount of capital provided by VC investors may affect the intensity with which a venture internationalizes. I use *Total Funding*, the total dollar value of VC investments from 1997 to 2003, as a control. Using the natural log of this variable compensates for its positive skew. Repeated exposure to VC providers may help the business owner internalize the venture capitalist's knowledge. I thus use the variable *Total Rounds,* which measures the number of VC investment rounds through 2003, as a control.

Manufacturers and service providers differ in their ability to export and the degree of capital required for international operations. Brock, Yaffe and Dembovsky (2006) suggest that contradictory results in previous studies of internationalization may have been because of sector differences: services and product companies differ in the potential for separability between producer and consumer, intangibility and nonfungibility of the offering, and high fixed-to-variable cost ratio. Service industries tend to be highly dynamic and competitive, with value based on intangible resources (Hitt, Bierman, Uhlenbruck & Shimizu, 2006). Such industry differences suggest variance in the international intensity of product and services. I thus use the dummy variable *Product/Service,* coded as "1" if the primary SIC of the venture indicates manufacturing and "0" otherwise.

Finally, the receipt of VC from foreign sources may provide some of the same market knowledge benefits that accrue from internationally experienced domestic CVC providers. Thus, I include the control variable *Foreign VC,* a dummy variable taking a value of "1" if the focal firm ever received VC from a foreign-based organization and "0" otherwise.

Analysis

Ventures selecting an internationalization strategy may have certain characteristics that enhance the probability of high international intensity.

Solely domestic ventures may have those same characteristics; thus, to determine the relationship between those characteristics and international intensity, one must address the selection bias. I do so by utilizing a Heckman two-stage procedure that employs a selection model and an effects model.

The selection model uses the dependent variable *International.* This variable is coded “1” if the focal firm ever had international sales and “0” otherwise. Independent variables in the selection model address factors associated with a venture conducting international business, but not necessarily with the intensity with which it conducts that business. The first of these, *Size,* is the number of employees the venture had at the reported date closest to the date of the 2003 sales data. International expansion requires a resource commitment that larger firms may better support (Aaby & Slater, 1989). Once a firm has initiated foreign activities, the intensity of those activities is not size dependent (Calof, 1993). I use the number of employees as a measure of size because many young firms either do not have sales or are early in their sales growth (Maula & Murray, 2001). Prior research of venture-backed firms (e.g., George, et al., 2005) has used this measure of size. I use the natural log of the number of employees because of the skew in the raw data.[12]

A venture’s stage of development may affect its decision to internationalize. Companies in early stages are still developing and testing the commercial viability of their offering, which may limit their ability to conduct international sales, whereas those in middle or late stages are growing and expanding (McNally, 1995). The variable *Early Stage* reflects the stage of the venture at its last round of venture financing prior to the reported 2003 date of sales and conforms to Gompers (1995). The dummy variable takes a value of “1” if the venture is in an early stage (“early” or “seed” per SDC), when international sales are less likely or “0” otherwise.

The final independent variable in the selection equation relates to industry. Historically, VC investments have targeted firms in “high-technology” industries, including communications, computers, electronics, biotech and medical/health. Although heterogeneity exists across industries, Gompers (1995) found that high-tech firms always receive at least 70 percent of the rounds disbursed in a given year. Industries differ in the appropriability of their intellectual property and in their approach to its protection (Levin, Cohen & Mowery, 1985). Industries may thus vary in the use of CVC, due to the risks of appropriation of such intellectual property by corporate investors (Keil, Zahra & Maula, 2004). In addition, high knowledge intensity associated with certain industries relates to faster internationalization (Autio et al., 2000), although licensing and regulatory issues may delay internationalization in some industries. To capture cross-industry differences as completely as possible, I included a set of dummy variables for CorpTech industry groups in the regressions. These variables are coded “1” for the industry of the focal venture and “0” otherwise. For model specification purposes (i.e., to ensure a good ratio of observations to variables), I did not include variables for industries that constituted less than 1 percent of the sample.

Table 1 presents a summary and descriptive statistics of the variables identified above.

The selection model is:

$$Probit\ (\Pr(International)) = \alpha + \beta_1\ Size + \beta_2\ Early\ Stage + \Sigma_m\ \beta_m\ Industry_m$$

where $\Sigma_m\ \beta_m\ Industry_m$ is a vector of industry dummies.

The general effects model is:

$$Probit\ (\Pr(High\ Intensity)) = \alpha + \beta_i\ CVC + \beta_j\ CVC * Moderator_j + \Sigma_m\ \beta_m\ Control_m$$

where *Moderator*$_j$ is one of the moderator variables identified above (*CVC Experience, CVC Presence, Age 1st CVC, CVC Rounds*), *CVC* * *Moderator*$_j$ is the interaction effect of the moderator with *CVC*, and $\Sigma_m\ \beta_m\ Control_m$ is a vector of the control variables.

Table 1. Variables: Sources and Descriptive Statistics

Variable	Definition	Source	Mean	Std.Dev.	Min	Max
Effects Model						
High Intensity	Dummy variable: '1' if 2003 ratio of foreign sales to total sales (FSTS) >10% (0 otherwise)	CorpTech	0.239	0.427	0	1
CVC	Dummy variable: '1' if venture received VC from a corporation (Fund type = CORPVEN) 1997-2003 (0 otherwise)	SDC	0.470	0.500	0	1
CVC Experience	Number of investments that corporate investors in the focal firm had provided to foreign ventures 1994-2002	SDC	0.519	1.191	0	10
CVC Presence	Number of countries in which corporate investors in the focal venture had subsidiaries or affiliates in 2001	Dir. of Corporate Affiliations, 2001	4.556	11.196	0	79
Age 1st CVC	The age, in months, at which the focal company first received venture capital from a corporation	SDC	10.830	15.543	-2.005	69.271
CVC Rounds	Number of rounds of venture funding in which corporate investors participated for the focal venture	SDC	0.728	0.977	0	5
Age	Age in years of the venture as of the 2003 sales date	SDC & CorpTech	4.609	1.455	0.068	6.995
Total Funding	Total amount of VC received by the focal venture 1997-2003.	SDC	9.842	1.199	5.753	13.112
Rounds	Number of rounds of VC received by the focal venture 1997-2003	SDC	3.403	1.801	1	11
Foreign VC	Dummy variable: '1' if venture received VC from foreign entity 1997-2003.	SDC	0.310	0.463	0	1
Product/ Service	Dummy variable: '1' if the primary SIC of the venture is in Division D: Manufacturing (0 otherwise).	CorpTech	0.317	0.466	0	1

Table 1. (Continued)

Variable	Definition	Source	Mean	Std.Dev.	Min	Max
Selection Model						
International	Dummy variable: '1' if the venture reported interntional sales in period 1997-2003 (0 otherwise)	CorpTech	0.519	0.501	0	1
Early Stage	Dummy variable: '1' if the stage of the venture at the latest funding round in the period 1997-2003 was "early" (seed or startup), (0 otherwise)	SDC	0.172	0.378	0	1
Size	Number of employees of focal venture at reporting date closest to end of 2003	CorpTech	3.627	1.119	0	6.686
Biotechnology	Dummy variable: '1' if the venture's primary industry is Biotechnology (0 otherwise)	CorpTech	0.078	0.269	0	1
Chemicals	Dummy variable: '1' if the venture's primary industry is Chemicals (0 otherwise)	CorpTech	0.022	0.148	0	1
Computer Hardware	Dummy variable: '1' if the venture's primary industry is Computer Hardware (0 otherwise)	CorpTech	0.037	0.190	0	1
Computer Software	Dummy variable: '1' if the venture's primary industry is Computer Software (0 otherwise)	CorpTech	0.381	0.486	0	1
Medical	Dummy variable: '1' if the venture's primary industry is Medical (0 otherwise)	CorpTech	0.034	0.180	0	1
Pharmaceuticals	Dummy variable: '1' if the venture's primary industry is Pharmaceuticals (0 otherwise)	CorpTech	0.056	0.230	0	1
Photonics	Dummy variable: '1' if the venture's primary industry is Photonics (0 otherwise)	CorpTech	0.026	0.160	0	1
Telecomm & Internet	Dummy variable: '1' if the venture's primary industry is Telecommunications and Internet (0 otherwise)	CorpTech	0.306	0.462	0	1

Table 2. Variable Correlations

	High Intensity	Inter-national	CVC	CVC Experience	CVC Presence	Age 1st CVC	CVC Rounds	Age	Total Funding	Rounds	Foreign VC	Product/ Service	Early Stage	Size
High Intensity	1.000													
Inter-national	0.540	1.000												
CVC	0.139	0.070	1.000											
CVC Experience	0.094	0.037	0.463	1.000										
CVC Presence	0.042	-0.053	0.433	0.430	1.000									
Age 1st CVC	0.145	0.089	0.741	0.237	0.361	1.000								
CVC Rounds	0.165	0.076	0.792	0.694	0.444	0.471	1.000							
Age	0.237	0.370	0.153	0.124	0.095	0.279	0.174	1.000						
Total Funding	0.022	-0.023	0.353	0.313	0.291	0.203	0.342	0.140	1.000					
Rounds	-0.087	0.087	0.184	0.129	0.161	0.094	0.186	0.230	0.478	1.000				
Foreign VC	0.022	-0.049	0.226	0.224	0.089	0.177	0.195	0.047	0.303	0.178	1.000			
Product/ Service	0.032	-0.146	0.033	-0.001	0.072	0.045	0.010	-0.129	0.010	0.039	0.116	1.000		

Table 2. (Continued)

	High Intensity	Inter-national	CVC	CVC Experience	CVC Presence	Age 1st CVC	CVC Rounds	Age	Total Funding	Rounds	Foreign VC	Product/ Service	Early Stage	Size
Early Stage	0.070	-0.037	-0.112	-0.132	-0.160	-0.141	-0.147	-0.284	-0.388	-0.366	-0.048	0.009	1.000	
Size	0.062	0.074	0.189	0.222	0.152	0.113	0.184	0.142	0.549	0.183	0.148	-0.152	-0.215	1.000

	Biotech	Chem	Computer Hardware	Medical	Pharma	Photonics	Computer Software	Telecomm & Internet
Bio-technology	1.000							
Chemicals	-0.044	1.000						
Computer Hardware	-0.057	-0.030	1.000					
Medical	-0.054	-0.028	-0.037	1.000				
Pharma-ceuticals	-0.071	-0.037	-0.048	-0.045	1.000			
Photonics	-0.048	-0.025	-0.032	-0.031	-0.040	1.000		
Computer Software	-0.229	-0.119	-0.154	-0.146	-0.191	-0.128	1.000	
Telecomm & Internet	-0.194	-0.101	-0.131	-0.124	-0.162	-0.109	-0.521	1.000

Analyses And Results

As seen in Table 1, slightly fewer than half (47 percent) of the sample companies have received CVC from US-based corporations. The average number of VC rounds received from inception through 2003 is 3.4, with a maximum of 11 rounds. Sample companies that received CVC obtained their first round of venture capital at an average age of six months. Companies were relatively small, averaging 36.7 employees; all but two of the ventures had 500 or fewer employees. The companies represent 33 different 4- digit SIC codes and 14 CorpTech primary industries.

Table 2 presents the correlations for the study variables. A review of the correlations concludes that few variables have correlations greater than 0.6, suggesting that multi-colinearity is generally not an issue in the analyses. High correlations are present between three pairs of variables relating to ever having received CVC, age at first CVC, and number of CVC rounds.

Companies in the sample that have received CVC (N=126) and those that have not (N=142) differ in a number of respects, as seen in Table 3. On average, they are further along in their development (13 percent early stage vs. 21 percent), larger (79 vs. 53 employees), and about five months older. In addition, they have received more than twice the venture funding in more rounds than ventures that have not received CVC, and are twice as likely to have received VC from foreign sources. T-tests determined the significance of differences in the means between the two groups; all the differences noted above are statistically significant. In terms of intensity of internationalization, 30 percent of ventures receiving CVC had a high level of international intensity, whereas only 18 percent of those not receiving CVC had a similarly high level of FSTS.

The sample companies received funding from 116 identified U.S. corporations and 452 identified independent venture capital providers. An additional 144 identified investors included 60 financial corporations, 48 investment banks and 11 individuals. I have provided a list of the U.S, corporate investors in Appendix B.

A similar comparison of international and domestic new ventures (also seen in Table 3) indicates that the former are, on average, one year older and more often in service industries. Industry differences are apparent: ventures in health sciences (biotechnology, medical and pharmaceuticals) are more likely to be domestic, whereas those in information technology (computer software and telecommunications & Internet) are more likely to be international.

Table 3. Comparison of ventures by CVC backing and Internationalization

	Received US CVC			International		
	No (N=142)	Yes (N=126)		No (N=129)	Yes (N=139)	
High Intensity	**0.18**	**0.30**	*	**0.00**	**0.46**	**
	(0.39)	(0.46)		0.00	(0.50)	
International	**0.49**	**0.56**		**0.00**	**1.00**	
	(0.50)	(0.50)		0.00	0.00	
CVC	**0.00**	**1.00**		**0.43**	**0.50**	
	0.00	0.00		(0.50)	(0.50)	
CVC Experience	**0.00**	**1.10**	***	**0.47**	**0.56**	
	0.00	(1.54)		(1.10)	(1.27)	
CVC Presence	**0.00**	**9.69**	***	**5.17**	**3.99**	
	0.00	(14.75)		(12.93)	(9.32)	
Age 1st CVC	**0.00**	**23.04**	***	**9.39**	**12.16**	
	0.00	(15.25)		(15.13)	(15.86)	
CVC Rounds	**0.00**	**1.55**	***	**0.65**	**0.80**	
	0.00	(0.87)		(0.92)	(1.03)	
Age	**4.40**	**4.84**		**4.05**	**5.13**	**
	(1.56)	(1.29)		(1.48)	(1.22)	
Total Funding	**23389.13**	**47896.21**	***	**37693.37**	**32329.02**	
	(32163.8 8)	(58343.38)		(56939.92)	(37545.71)	
Rounds	**3.09**	**3.75**	***	**3.24**	**3.55**	
	(1.82)	(1.72)		(1.57)	(1.99)	
Foreign VC	**0.21**	**0.42**	***	**0.33**	**0.29**	
	(0.41)	(0.50)		(0.47)	(0.45)	
Product/ Service	**0.30**	**0.33**		**0.33**	**0.25**	*
	(0.46)	(0.47)		(0.49)	(0.44)	
Early Stage	**0.21**	**0.13**	+	**0.19**	**0.16**	
	(0.41)	(0.33)		(0.39)	(0.37)	
Size	**53.09**	**78.93**	*	**63.81**	**66.56**	
	(78.98)	(102.80)		(107.28)	(74.76)	
Biotechnology	**0.08**	**0.08**		**0.13**	**0.03**	**
	(0.27)	(0.27)		(0.34)	(0.17)	
Chemicals	**0.02**	**0.02**		**0.02**	**0.02**	
	(0.14)	(0.15)		(0.15)	(0.15)	
Computer Hardware	**0.03**	**0.05**		**(0.03)**	**0.04**	
	(0.17)	(0.21)		(0.17)	(0.20)	
Medical	**0.04**	**0.02**		**0.03**	**0.01**	*
	(0.20)	(0.15)		(0.24)	(0.08)	
Pharmaceuticals	**0.06**	**0.06**		**0.10**	**0.01**	**
	(0.23)	(0.23)		(0.30)	(0.12)	

Table 3. (Continued)

	Received US CVC			International		
	No (N=142)	Yes (N=126)		No (N=129)	Yes (N=139)	
Photonics	**0.02**	**0.03**		**0.02**	**0.03**	
	(0.14)	(0.18)		(0.15)	(0.17)	
Computer Software	**0.40**	**0.36**		**0.29**	**0.46**	**
	(0.49)	(0.48)		(0.46)	(0.50)	
Telecomm & Internet	**0.29**	**0.33**		**0.25**	**0.36**	*
	(0.45)	(0.47)		(0.43)	(0.48)	

The Heckman procedure reports coefficients for both the selection model and the effects model as seen in Table 4. Significant F values and Rho statistics (a measure of correlation between the models) confirm that the chosen two-stage Heckman regressions were appropriate.

I begin with some comments on the selection models before presenting the results of the six effects models. The selection models are consistent in that neither the stage of the venture nor its size has a significant relationship with having had international activity. However, certain industries do have such relationships. Specifically, ventures in information technology industries (computer hardware, computer software and telecommunications & Internet) have higher probabilities of having had international activities. Ventures in these industries are 2.2 to 2.5 times more likely to have had international sales than ventures in other industries.

The results of Model 1 indicate a positive and significant relationship between receipt of CVC and increased chances of a high level of international intensity after controlling for factors likely to affect a high degree of international intensity. Specifically, the odds of a CVC-backed venture deriving more than 10 percent of its revenue from foreign sources are almost 60 percent greater than the odds for a non-CVCbacked venture.[13] In other words, in a sample of 100 companies, half of which received CVC, the odds are that 16 of those receiving CVC have high international intensity compared with only 10 of the non-CVC ventures. This supports Hypothesis 1 that a positive relationship exists between receiving CVC and high levels of international intensity. I believe this indicates a relationship of great economic importance.

Table 4. Logistic Regression Analysis Results

	(1)	(2)	(3)	(4)	(5)	(6)
Effects Model						
CVC	0.463*	0.425+	0.401+	0.46	0.179	-0.006
	(0.208)	(0.220)	(0.219)	(0.315)	(0.297)	(0.449)
CVC x CVC Experience		0.039				-0.064
		(0.080)				(0.112)
CVC x CVC Presence			0.009			0.007
			(0.010)			(0.012)
CVC x Age 1st CVC				0		0.004
				(0.010)		(0.011)
CVC x CVC Rounds					0.186	0.248
					(0.148)	(0.204)
Age	0.039	0.036	0.038	0.039	0.035	0.024
	(0.078)	(0.078)	(0.078)	(0.085)	(0.079)	(0.087)
Total Funding	0.113	0.108	0.1	0.113	0.103	0.097
	(0.101)	(0.102)	(0.102)	(0.101)	(0.102)	(0.104)
Rounds	-0.204**	-0.204**	-0.203**	-0.204**	-0.210**	-0.208**
	(0.075)	(0.075)	(0.075)	(0.075)	(0.077)	(0.079)
Foreign VC	0.211	0.2	0.222	0.211	0.205	0.229
	(0.213)	(0.215)	(0.214)	(0.213)	(0.215)	(0.218)
Product/ Service	0.524*	0.528*	0.526+	0.524*	0.544*	0.547+
	(0.266)	(0.267)	(0.269)	(0.266)	(0.276)	(0.281)
Constant	-1.633+	-1.574+	-1.511	-1.631+	-1.509	-1.403
	(0.930)	(0.937)	(0.941)	(0.940)	(0.938)	(0.968)
Selection Model						
Early Stage	-0.149	-0.153	-0.157	-0.149	-0.163	-0.168
	(0.220)	(0.221)	(0.222)	(0.220)	(0.224)	(0.228)
Size	0.002	0.003	0.003	0.002	0.001	0.001
	(0.074)	(0.074)	(0.074)	(0.074)	(0.075)	(0.075)
Bio-technology	-0.218	-0.214	-0.228	-0.218	-0.236	-0.257
	(0.435)	(0.435)	(0.434)	(0.435)	(0.434)	(0.434)
Chemicals	0.607	0.61	0.615	0.606	0.568	0.544
	(0.574)	(0.574)	(0.572)	(0.575)	(0.575)	(0.581)
Computer Hardware	0.804+	0.810+	0.779	0.803+	0.801+	0.76
	(0.486)	(0.486)	(0.487)	(0.487)	(0.483)	(0.491)
Medical	-0.687	-0.692	-0.707	-0.687	-0.734	-0.739
	(0.626)	(0.626)	(0.626)	(0.627)	(0.626)	(0.627)
Pharma-ceuticals	-0.35	-0.35	-0.363	-0.35	-0.384	-0.407
	(0.500)	(0.500)	(0.499)	(0.500)	(0.500)	(0.500)
Photonics	0.573	0.575	0.564	0.573	0.533	0.508
	(0.561)	(0.562)	(0.562)	(0.562)	(0.568)	(0.574)
Computer Software	0.896**	0.900**	0.891**	0.896**	0.867**	0.853*
	(0.337)	(0.337)	(0.335)	(0.337)	(0.336)	(0.337)

Table 4. (Continued)

	(1)	(2)	(3)	(4)	(5)	(6)
Telecomm & Internet	0.932**	0.929**	0.91 6**	0.932**	0.897**	0.880**
	(0.340)	(0.340)	(0.339)	(0.340)	(0.338)	(0.339)
Constant	-0.586	-0.589	-0.579	-0.586	-0.551	-0.532
	(0.407)	(0.407)	(0.406)	(0.407)	(0.410)	(0.412)
athrho: Constant	1.051+	1.053+	1.057+	1.051+	1.049	1.042
	(0.623)	(0.624)	(0.642)	(0.624)	(0.688)	(0.713)
Observations	268	268	268	268	268	268
Censored observations	139	139	139	139	139	139
Uncensored observations	129	129	129	129	129	129
Wald chi2 =	12.74	12.85	12.72	12.74	12.57	12.41
Prob > chi2 =	0.0474	0.0758	0.0792	0.0788	0.0833	0.2584

Standard errors in parentheses
+ significant at 10%; * significant at 5%; ** significant at 1%

Models 2 through 5 include interaction effects for the hypothesized moderator variables, with main effects excluded, as they are the same as the interaction effects due to the dichotomous nature of the variable *CVC*. None of the interactions is significant in these models. Model 2 fails to find support for the relationship between high international intensity and the prior international experience of the CVC unit proposed in Hypothesis 2. However, as in Model 1, the receipt of CVC remains significant and positive in its relationship with high international intensity, albeit with a slight decrease in the magnitude of the relationship (52 percent higher probability). Model 3 reports the same positive and significant relationship of a similar magnitude for CVC and high intensity but does not indicate support for the hypothesis that the foreign diversity of the corporation providing capital influences that relationship. Model 4 indicates no support for Hypothesis 4, which states that the age at which a venture first receives CVC relates to high international intensity. Model 5 similarly indicates no support for the hypothesis that the number of CVC rounds relates to high intensity. Finally, Model 6, which includes all moderator variables, is not significant.

The effects models also provide additional insights into other factors related to high international intensity. The models indicate a consistent significant relationship between the total number of VC rounds received and high intensity. In all six models, the coefficient of *Rounds* is highly significant and indicates that the marginal decrease in the probability of high intensity for each VC round is roughly 20 percent. In other words, each additional round of financing actually decreases the odds of high international intensity by about 18.5 percent.

An intriguing result was found in the relationship between whether the venture is in a product or service industry and high intensity. All six models show a significant and positive relationship between high intensity and operating in a product industry. The odds for a product venture having high intensity are almost 70 percent greater than those for a service venture. Whereas Table 4 indicated that product ventures are, on average, more often solely domestic, it appears that if they do internationalize, they are much more likely to do so intensively.

DISCUSSION

The finding of a positive relationship between receipt of corporate venture capital and a high level of international intensity is analytically and economically significant. The difference in the probabilities of CVC- and non-CVC-backed ventures achieving high internationalization may explain the contradictory findings of previous studies of venture capital and internationalization (Burgel & Murray, 1998; Carpenter et al., 2003). Those studies did not consider the differential value-added of alternate VC sources. The consistent magnitude of the relationship in four of the models suggests a degree of robustness to the relationship in the presence of various factors. The rather uniform nature of the coefficients of the controls across the six models underscores the robustness of the relationships. In other words, it seems clear that there is a significant relationship between CVC receipt and high international intensity, although the underlying mechanism by which this relationship exists is not clear in this study.

The finding that each additional financing round received by an INV is associated with a lower probability of high internationalization may suggest the presence of agency issues. VCs may provide internationalized ventures with more and smaller financing rounds as a means to mitigate agency problems attendant with foreign expansion. Smaller increments of capital may constrain the ability of ventures to deepen adequately their international presence. More (and more frequent) rounds of VC may permit investors to monitor the performance of an internationalized venture more closely, potentially enhancing its domestic performance and thus lowering its FSTS.

The lack of support for the prior international experience of the CVC provider may be due to the nature of the knowledge relevant for foreign intensity. Since I base the measure of CVC prior experience on investments in foreign firms and not on prior investments in U.S.-based INVs, it may have missed the more nuanced knowledge acquired by the CVC provider regarding internationalization of ventures, rather than acquired market knowledge. Both are tacit and hard to measure.

The relationship between industry and internationalization warrants further study. The analyses show that certain industries (e.g., information technology) are associated with having had international sales. More ventures in product industries are, on average, solely domestic, yet product companies are more likely to have high international intensity. It may be that once such ventures pass the hurdles to international sales (e.g., regulatory barriers, plant/equipment capital expenditures), they may more easily scale their international operations.

Contributions and Implications

This paper's principal contribution may be to theory development regarding INVs, since many portfolio firms receiving early-stage venture capital are young. The INV literature utilizes various cutoff levels of FSTS for sample selection, from as low as 5 percent to as high as 25 percent, with a consensus developing around 10 percent (Coviello & Jones, 2004). This paper shows a significant difference between firms above this threshold and those below it—in this case, it is the receipt of CVC. Extensions of the current study will utilize different cutoff levels for high international intensity (e.g., 25 percent, the level used by Knight & Cavusgil, 2005) to help provide empirical support for the consensus. The discovery/determination of a relationship between CVC and international intensity will clearly benefit INV theory development.

Regarding empirical findings, this study extends work by Maula et al. (2005) of the differences in value between independent and corporate venture capital in the international domain. Whereas their study examined perceived differences in value-added because of the funding source, the current study empirically examines outcomes related to use of different funding sources. This study validates the perception that CVC providers offer more value in supporting international business than do IVCs. This study also extends Carpenter et al. (2003) by the contrary finding—a relationship between the receipt of some types of CVC and greater international intensity. The value-added of CVC may be responsible for the different finding.

APPENDIX A. PROBIT ANALYSIS RESULTS

	(1)	(2)	(3)	(4)	(5)	(6)
CVC	0.351+	0.325	0.369+	0.461+	0.085	0.099
	(0.190)	(0.204)	(0.204)	(0.278)	(0.283)	(0.404)
CVC x CVC Experience		0.029				-0.045
		(0.079)				(0.103)
CVC x CVC Presence			-0.002			-0.004
			(0.009)			(0.010)
CVC x Age 1st CVC				-0.005		-0.001
				(0.009)		(0.009)
CVC x CVC Rounds					0.176	0.229
					(0.137)	(0.178)
Age	0.283**	0.281**	0.283**	0.295**	0.278**	0.282**
	(0.068)	(0.068)	(0.068)	(0.073)	(0.069)	(0.074)
Total Funding	0.052	0.048	0.055	0.047	0.044	0.053
	(0.092)	(0.093)	(0.092)	(0.092)	(0.092)	(0.094)
Rounds	-0.171**	-0.171**	-0.170**	-0.173**	-0.177**	-0.178**
	(0.066)	(0.066)	(0.066)	(0.066)	(0.067)	(0.067)
Foreign VC	-0.002	-0.01	-0.006	0.006	-0.006	0.003
	(0.199)	(0.200)	(0.199)	(0.199)	(0.200)	(0.203)
Product/ Service	0.231	0.233	0.233	0.238	0.242	0.248
	(0.190)	(0.190)	(0.191)	(0.191)	(0.191)	(0.192)
Constant	-2.267**	-2.215*	-2.295**	-2.272**	-2.156*	-2.254**
	(0.849)	(0.861)	(0.857)	(0.849)	(0.855)	(0.868)
Observations	268	268	268	268	268	268

Standard errors in parentheses
+ significant at 10%; * significant at 5%; ** significant at 1%

For practitioners, this study helps inform entrepreneurs as to types of equity capital to obtain given goals of internationalization. While VC is rare, those firms that obtain VC receive multiple offers (Smith, 2001), and therefore should make selections based on strategic objectives. For example, ventures following an internationalization strategy may seek to obtain resources such as CVC that support that strategy. Ventures following domestic strategies may make VC selection decisions based on their strategic objectives. Venture capitalists consider the human capital of the entrepreneurial team in funding decisions (Smart, 1999), but their decision to fund changes the human capital of the portfolio company by their involvement. An ex ante understanding of the respective human and social capital the entrepreneur and the venture capital firm each bring to the relationship may change the costs of the funding. Further, identifying which aspects of venture capital lead to high international intensity and how those factors vary from domestic settings may change the way that entrepreneurs select venture capital providers, the expectations they have of their relationship with them, and the way they attempt to utilize them.

Finally, understanding the factors that moderate the relationship between receipt of CVC and international intensity further informs policymakers. For example, if CVC investments at earlier ages are associated with higher rates of international intensity, policymakers may create investment incentives for multinationals that integrate that factor. This may have significant implications in emerging economies seeking to develop nascent industries.

Appendix B. U.S. Corporate Venture Capital Providers to Sample Ventures

@Ventures
Abbott Laboratories
Accenture Technology Ventures (FKA: AC Ventures)
Affymetrix, Inc.
Alliance Venture Management, LLC
Altera Corporation
Ameritech Development Corp.
Anschutz Investment Company
BD Ventures (AKA: Becton, Dickinson & Co.)
BEA Systems
BlueVector, LLC
Brooks Fiber Properties, Inc.
Cargill Ventures
Caterpillar Venture Capital, Inc.
ChevronTexaco Technology Ventures LLC (CTTV Investments)
Child Health Investment Company, LLC
Cinergy Ventures
Cisco Systems, Inc.
Comcast Interactive Capital
Comdisco Ventures
Conexant Systems, Inc.

Corning Innovation Ventures
Cox Enterprises, Inc.
Cypress Semiconductor Corporation (FKA Intjr, Inc.)
Dell Ventures
Dow Chemical Company, The
Duchossois Technology Partners LLC (DTEC)
Eastman Kodak Co, Inc.
Eastman Ventures
eBay Inc.
Exelon Capital Partners
Fenwick & West LLP
Fletcher Spaght Associates
Garage Technology Ventures (FKA: Garage.com)
GE Equity (FKA: GE Capital Equity Capital Group)
Genentech Corporation
Global Crossing Ventures (FKA: Frontier Ventures)
Gray Cary Ware & Freidenrich
Hearst Corporation
Hewlett-Packard Strategy and Corporate Development
Hillman Company
Hollinger International, Inc.
Honeywell Inc.
iNCUBIC
Intel Capital
internet.com
Jefferies Group, Inc.
Johnson & Johnson Development Corporation
Juniper Networks
Kyocera International, Inc.
Lilly BioVentures
Lilly Ventures (FKA: e.Lilly Ventures)
Lotus Development Corporation
Lucent Venture Partners, Inc.
Lycos Ventures
Marcus & Millichap Venture Partners (AKA: MMVP)
Matsushita Electric Corporation of America
MC Capital, Inc.
MedImmune

Merck Capital Ventures
Microsoft Corporation
Motorola Ventures
NEC USA, Inc.
Novartis Corp.
Office Depot, Inc.
Opticality Ventures
Oracle Venture Fund
OrbiMed Advisors LLC
Pacific Venture Capital, LLC
Plantronics Inc.
POSCO BioVentures
Procter & Gamble Company, The
Proctor & Gamble
PSINet Ventures
R.R. Donnelley & Sons
Rare Ventures
RCT BioVentures NE LLC
RSA Capital
Rustic Canyon Partners(FKA: TMCT Ventures, L.P.)
SAIC Venture Capital Corporation
Siemens Corporation
Summit Design, Inc.
Sun Microsystems, Inc.
Teradyne
Total Technology Ventures LLC (AKA: TTV)
Tribune Ventures
Village Ventures
Visa International
Vitesse Semiconductor Corporation
Williams Communications Group
Wilson, Sonsini, Goodrich & Rosati (AKA: WS Investments)
Wind River Ventures
WorldCom Venture

APPENDIX C. ACRONYMS

CVC	Corporate venture capital
FSTS	Ratio of foreign sales to total sales
INV	International new venture
INVs	International new ventures
IVC	Independent venture capital
MNC	Multinational corporation
MNE	Multinational enterprise
PC	Portfolio company
PCs	Portfolio companies
SDC	Ventrure Xpert/SDC Platinum
TMT	Top management team
VC	Venture capital

BIBLIOGRAPHY

Aaby, N. E. & Slater, S. F. (1989). "Management influences on export performance: A review of the empirical literature 1978-88." *International Marketing Review, 6*(4), 7-26.

Alvarez, S. A. & Busenitz, L. W. (2001). "The entrepreneurship of resource-based theory." *Journal of Management, 27,* 755-775.

Athanassiou, N. & Nigh, D. (2000). "Internationalization, Tacit Knowledge and the Top Management Teams of MNCs." *Journal of International Business Studies*, *31*(3), 471- 487.

Autio, E., & Sapienza, H. et al. (2000). "Effects of age at entry, knowledge intensity, and imitability on international growth." *Academy of Management Journal* , *43*(5), 909-924.

Barney, J. (1991). "Firm resources and sustained competitive advantage." *Journal of Management,* 17, 99-121.

Bloodgood, J. M., & Sapienza, H. J. et al. (1996). "The internationalization of new high- potential U.S. ventures: Antecedents and outcomes." *Entrepreneurship Theory & Practice,* Summer, 1996, 61-76.

Brock, D. M., & Yaffe, T. et al. (2006). "International diversification and performance: A study of global law firms." *Journal of International Management, 12*, 473-489.

Brush, C. G. (1992). Factors motivating small firms to internationalize: the effect of firm age. *Unpublished doctoral dissertation.* Boston, Boston University.

Brush, C. G. (1995). *International entrepreneurship: The effect of firm age on motives for internationalization.* New York, Garland Publishing, Inc.

Burgel, O. & Murray, G. C. (1998). *The international activities of british start-up companies in high-technology industries: Differences between internationalisers and non-internationalisers.* Babson Kauffman Entrepreneurship Research Conference, Babson College.

Calof, J. L. (1993). "The impact of size on internationalization." *Journal of Small Business Management, 31*(4), 60-69.

Carpenter, M. A. & Pollock, T. G. et al. (2003). "Testing a model of reasoned risk-taking: Governance, the experience of principals and agents, and global strategy in high- technology IPO firms." *Strategic Management Journal, 24*(9), 803-820.

Carpenter, M. A. & Sanders, W. G. (2004). "The effects of top management team pay and firm internationalization on MNC performance." *Journal of Management, 30*(4), 509- 528.

Caves, R. E. (1996). *Multinational enterprise and economic analysis.* Cambridge, MA, Cambridge University Press.

Chen, R. R. & Martin, M. J. (2001). "Foreign expansion of small firms: The impact of domestic alternatives and prior foreign business involvement." *Journal of Business Venturing, 16*, 557-574.

Cohen, W. & Levinthal, D. (1990). "Absorptive capacity: A new perspective on learning and innovation." *Administrative Science Quarterly, 35*, 128-152.

Coviello, N. E. & Jones, M. V. (2004). "Methodological issues in international entrepreneurship research." *Journal of Business Venturing, 19*, 485-508.

Coviello, N. E. & McAuley, A. (1999). "Internationalisation and the smaller firm: A review of contemporary empirical research." *Management International Review, 39*, 223- 256.

Coviello, N. E. & Munro, H. J. (1995). "Growing the entrepreneurial firm: Networking for international market development." *European Journal of Marketing,* 29(7), 49ff. *Directory of corporate affiliations*, National Register Publishing. 2001

Daily, C. M., & Certo, S. T. et al. (2000). "International experience in the executive suite: The path to prosperity?" *Strategic Management Journal,* 21, 5 15-523.

Dunning, J. H. (1988). "The eclectic paradigm of international production: A restatement and some possible extensions." *Journal of International Business Studies, 19*, (Spring 1988), 1-30.

Erramilli, M. K. & Rao, C. P. (1990). "Choice of foreign market entry modes by service firms: Role of market knowledge." *Management International Review, 30*, 135-150.

Evans, D. S. & Jovanovic, B. (1989). "An estimated model of entrepreneurial choice under liquidity constraints." *Journal of Political Economy, 97*(4), 808-826.

George, G., Wiklund, J. et al. (2005). "Ownership and the internationalization of small firms." *Journal of Management, 31*(2), 2 10-233.

Gompers, P. A. (1995). "Optimal investment, monitoring, and the staging of venture capital." *Journal of Finance, 50*, 1461-1490.

Gompers, P.A. & Lerner, J. (1998). The determinants of corporate venture capital success: Organizational structure, incentives and complementarities. *NBER #6725.* Cambridge, MA, NBER #6725.

Gompers, P. A. & Lerner, J. (1999). *The venture capital cycle.* Cambridge, MIT Press.

Gompers, P. A. & Lerner, J. (2000). "Money chasing deals? The impact of fund inflows on the valuation of private equity investments." *Journal of Financial Economics,* 55, 281- 325.

Gorman, M. & Sahlman, W. A. (1989). "What do venture capitalists do?" *Journal of Business Venturing,* 4(4), 231-248.

Grant, R. M. (1996). "Toward a knowledge-based theory of the firm." *Strategic Management Journal, 17* (Winter Special Issue), 109-122.

Green, M. B. (2004). "Venture capital investment in the United States 1995-2002." *The Industrial Geographer, 2*(1), 2-30.

Guler, I. & Guillen, M. F. (2005). *Knowledge, institutions, and foreign entry: The internationalization of U.S. venture capital firms*, Boston University and The Wharton School.

Gupta, A. K. & Sapienza, H. J. (1992). "Determinants of venture capital firms' preferences regarding the industry diversity and geographic scope of their investments." *Journal of Business Venturing, 7*(3), 347-362.

Harveston, P. D. & Kedia, B. L. et al. (2000). "Internationalization of born global and gradual globalizing firms: The impact of the manager." *Advances in Competitiveness Research, 8*(1), 92-99.

Heckman, J. J. (1979). "Sample selection bias as a specification error." *Econometrica, 47*(1), 153-161.

Hellmann, T. & Puri, M. (2002). "Venture capital and the professionalization of start-up firms: Empirical evidence." *The Journal of Finance,* LVII(1), 169-197.

Hitt, M. A. & Bierman, L. et al. (2006). "The importance of resources in the internationalization of professional service firms: The good, the bad, and the ugly." *Academy of Management Journal, 49*(6), 1137-1157.

Hymer, S. H. (1976). *TheiInternational operations of national firms: A study of direct foreign investment.* Cambridge, Massachusetts, The MIT Press.

International Trade Administration, *Small & medium-sized exporting companies: statistical overview, 2003*, Office of Trade and Industry Information, International Trade Administration, U.S. Department of Commerce.

Johanson, J. & Vahlne, J. (1977). "The internationalization process of the firm: a model of knowledge development and increasing foreign commitments." *Journal of International Business Studies, 8*(1), 23-32.

Jones, M. V. (1999). "Theinternationalization of small high-technology firms." *Journal of International Marketing, 7*(4), 15-41.

Kann, A. (2001). *Strategic venture capital investing by corporations: A framework for structuring and valuing corporate venture capital programs,* Stanford University.

Keil, T. & Zahra, S. A. et al. (2004). "Explorative and exploitative learning from corporate venture capital: A model of program level determinants." Academy of Management Best Conference Paper 2004.

Kelley, D. & Spinelli, S. (2001). T*he role of corporate investor relationships in the formation of alliances for corporate venture capital funded startups.* Babson Kauffman Entrepreneurship Research Conference.

Knight, G. A. & Cavusgil, S. T. (2004). "Innovation, organizational capabilities, and the born-global firm." *Journal of International Business Studies, 35*(2), 124-141.

Knight, G. A. & Cavusgil, S. T. (2005). "A taxonomy of born-global firms." *Management International Review, 45*(3), 15-35.

Knight, G. A. & Liesch, P. W. (2002). "Information internalisation in internationalising the firm." *Journal of Business Research, 55*(12), 981-995.

Kogut, B. & Zander, U. (1992). "Knowledge of the firm, combinative capabilities, and the replication of technology." *Organization Science 3*(3): 383-397.

Kogut, B. & Zander, U. (1993). "Knowledge of the firm and the evolutionary theory of the multinational corporation." *Journal of International Business Studies, 24*(4), 625-645.

Levin, R. C. & Cohen, W. M. et al. (1985). "R&D appropriability, opportunity, and market structure: New evidence on some Schumpeterian hypotheses." *American Economic Review, 75*, 20-24.

Lippman, S. & Rumelt, R. (1982). "Uncertain imitability: An analysis of interfirm differences in efficiency under competition." *Bell Journal of Economics (RAND Journal of Economics), 13*, 418-438.

Manigart, S. & Collewaert, V. et al. (2006). *Human capital and the internationalization of venture capital firms*, Vlerick Leuven Gent.

Maula, M. & Autio, E. et al. (2005). "Corporate venture capitalists and independent venture capitalists: What do they know, who do they know, and should entrepreneurs care?" *Venture Capita*l, *7*(1), 3-21.

Maula, M. & Murray, G. (2001). *Complementary value-adding roles of corporate venture capital and independent venture capital investors,* Helsinki University of Technology, Institute of Strategy and International Business.

Maula, M. & Murray, G. (2002). "Corporate venture capital and the creation of US public companies: The impact of sources of venture capital on the performance of portfolio companies." *Creating Value: Winners in the New Business Environment.* M. A. Hitt, R. Amit, C. E. Lucier and R. D. Nixon. Oxford, UK, Blackwell Publishing.

McDougall, P. P. & Oviatt, B. M. (2003). "Some fundamental issues in international entrepreneurship." *http://www.usasbe.org/knowledge/whitepapers/mcdougall2003.pdf.*

McDougall, P. P. & Oviatt, B. M. et al. (2003). "A comparison of international and domestic new ventures." *Journal of International Entrepreneurship, 1*(1), 59-82.

McNally, K. (1995). "Corporate venture capital: The financing of technology businesses." *International Journal of Entrepreneurial Behaviour & Research, 1*(3), 9-28.

McNally, K. (1997). C*orporate venture capital: Bridging the equity gap in the small business sector*. London, Routledge.

Oviatt, B. & McDougall, P. P. (1994). "Towards a theory of international new ventures." *Journal of International Business Studies, 25*(1), 45-63.

Oviatt, B. & McDougall, P. P. (1997). "Challenges for internationalization process theory: The case of international new ventures." *Management International Review, 37*(2), 85-99.

Oviatt, B. M. & McDougall, P. P. (2005). "Defining international entrepreneurship and modeling the speed of internationalization." *Entrepreneurship Theory & Practice.*

Oxley, J. & Sampson, R. C. (2003). "The scope and governance of international strategic alliances." *Strategy Management Journal, 25*(8/9), 723-749.

Preece, S. B. & Miles, G. et al. (1998). "Explaining the international intensity and global diversity of early-stage technology-based firms." *Journal of Business Venturing, 14*, 259- 281.

Rauser, v. I. (2002). *Value added of corporate venture capital: How do CVC units benefit from their organizational core?* Bamberg, Germany, Otto Friedrich University.

Reuber, A. R. & Fischer, E. (1997). "The influence of the management team's international experience on the internationalization behaviors of SMEs." *Journal of International Business Studies* Fourth Quarter (1997), 807-825.

Sapienza, H. (1992). "When do venture capitalists add value?" *Journal of Business Venturing, 9*(1), 9-27.

Sapienza, H. J. & Autio, E. et al. (2006). "A capabilities perspective on the effects of early internationalization on firm survival and growth." *Academy of Management Review, 31*(4), 914-933.

Sapienza, H. J. & Clercq, D. D. et al. (2005). "Antecedents of international and domestic learning effort." *Journal of Business Venturing, 20*(1), 437-457.

Shane, S. (2000). "Prior knowledge, and the discovery of entrepreneurial opportunities." *Organization Science* 11(4): 448-469.

Sharma, D. D. & Blomstermo, A. (2003). "The internationalization process of Born Globals: a network view." *International Business Review, 12*(6), 739-753.

Simonin, B. L. (1999). "Ambiguity and the process of knowledge transfer in strategic alliances." *Strategic Management Journal* 20(7): 595.

Smart, G. H. (1999). "Management assessment methods in venture capital: An empirical analysis of human capital valuation." *Venture Capital, 1*(1), 59-82.

Smith, G. (2001). "How early stage entrepreneurs evaluate venture capitalists." *Journal of Private Equity, 4*(2), 3 3-45.

Soh, P.-H. (2003). "The role of networking alliances in information acquisition and its implications for new product performance." *Journal of Business Venturing, 18*(6), 727- 744.

Sorenson, O. & Stuart, T. E. (2001). "Syndication networks and the spatial distribution of venture capital investments." *American Journal of Sociology, 106*(6), 1546-1588.

Stinchcombe, A. L. (1965). Social structure and organizations. *Handbook of Organizations*. J. G. March. Chicago, Rand-McNally: 142-193.

Stuart, T. E. & Hoang, H. et al. (1999). "Interorganizational endorsements and the performance of entrepreneurial ventures." *Administrative Science Quarterly, 44*, 315-349.

The state of small business (1992). *The U.S. Small Business Administration*, Washington, D.C., Government Printing Office (GPO).

Uzzi, B. (2000). "The sources and consequences of embeddedness for the economic performance of organizations." *American Sociological Review, 61*, 674-698.

Winter, S. (1987). "Knowledge and competence as strategic assets." *The Competitive Challenge: Strategies for Industrial Innovation and Renewal.* D. J. Teece. Cambridge, MA, Ballinger.

Yeoh, P. L. (2004). "International learning: antecedents and performance applications among newly internationalizing companies in an exporting context." *International Marketing Review, 21*(4/5), 511-535.

Zaheer, S. (1995). "Overcoming the liability of foreignness." *Academy of Management Journal, 38*, 341-363.

End Notes

[1] While not all small firms are young, most young firms are small, thus references to "small firms" qualified by age are appropriate to this study.

[2] For a list of U.S. corporate venture capital investors for this study, see Appendix B.

[3] http://www.census.gov/foreign-trade.

[4] Office of Advocacy, U.S. Small Business Administration, from data provided by the U.S. Bureau of the Census, Statistics of U.S. Business.

[5] Pratt's 1998 Guide to Venture Capital Sources.

[6] Securities Data Corporation, VentureXpert.

[7] In 2000, for example, over one-quarter of U.S. CVC deals were conducted by 12 CVC units in multinational companies such as Intel, General Electric, Dell, Cisco, Motorola, Sun, Compaq, H-P, PSINet, Microsoft and Andersen Consulting.

[8] Appendix A includes the result of a set of probit regressions that did not utilize a two-stage model. Coefficients in these models are markedly different in their significance and magnitude. In additional probit regressions (not shown), CVC was not significant.

[9] Funding limits precluded obtaining data for all years, thus my data includes bi-annual observations.

[10] SDC data show that 82% of the 38,204 rounds of venture capital disbursed between October 1995 and September 2005 (inclusive) went to companies that were six or fewer years old.

[11] Table 1 provides a summary of variables, their source and descriptive statistics.

[12] Because several respondents indicated having no employees, I added a value of one to this item prior to taking its log.

[13] In general, one interprets coefficients for variables in Heckman models as he/she would for any Logistic regression models; that is, one performs exponentiation on the coefficient.. In the current case, exp(.463) = 1.588.

In: The Role of Corporate Venture Capital in ... ISBN: 978-1-60876-936-0
Editors: Joseph F. Tollington

Chapter 2

CORPORATE VENTURE CAPITAL (CVC): SEEKING INNOVATION AND STRATEGIC GROWTH

Ian MacMillan, Edward Roberts, Val Livada and Andrew Wang
National Institute of Standards and Technology

ACKNOWLEDGMENTS

This study project arose from a common interest of industry practitioners, academic scholars, and government policymakers in the role of corporate venture capital (CVC) in technology innovation. The research included a review of academic literature and industry data, interviews with principals at corporate venture capital organizations, and collecting and analyzing survey data on corporate venture capital.

The main study team collaborated with the National Venture Capital Association (NVCA) to collect survey data from CVC organizations. The authors acknowledge the key roles played by Janice Mawson, Vice President of Membership, NVCA, in conducting outreach to CVC organizations and supporting the survey; Jeanne Metzger, Director of Marketing, NVCA, in design and implementation of the survey; and John Taylor, Vice President of Research, NVCA, in providing industry data on CVCs.

The authors acknowledge the role of Peter Holden, Partner, Coller Capital (formerly Research Fellow, Wharton School), in initiating the study project and providing contacts and insight during the effort.

The authors acknowledge Stephanie Shipp, Director, Impact Analysis Office, Technology Innovation Program, for her support and feedback throughout the study project; Robert Sienkiewicz, Supervisor, Impact Analysis Office, Technology Innovation Program, for his input and feedback in preparing this study report; and Lorel Wisniewski, Deputy Director, Technology Innovation Program, for her review of the final report.

ABSTRACT

This report examines corporate venture capital (CVC) as a model of innovation. CVC programs in established corporations invest in and partner with entrepreneurial companies. By doing so, established companies are able to identify and source new emerging technologies from entrepreneurial companies.

CVCs typically make a financial investment and receive a minority equity stake in an entrepreneurial company. CVCs also facilitate investment of in-kind resources into portfolio companies. In return, the parent corporation gains a window on new technologies and strategically complementary companies that could become strategic partners. CVCs generally invest with a combination of financial and strategic objectives. Strategic objectives include leveraging external sources of innovation, bringing new ideas and technologies into the company, and taking "real options" on technologies and business models (by investing in a wider array of technologies or business directions than the company can pursue itself).

Corporate venture capital may be viewed in the broader context of corporate venturing, including both internal and external venturing. Internal venturing programs "go inside" the firm and create entrepreneurial ventures from within the corporation. External venturing programs "go outside" the firm and tap external sources of innovation, whether through research collaborations with universities, strategic alliances with other firms, or partnerships with entrepreneurial companies. Often, the firm's internal and external venturing efforts are closely related and interact with each other.

CVC programs in established corporations face both inward and outward. They face outward to build relationships with the entrepreneurial venture community, learn about new technology and business directions, and make investments that create new strategic opportunities for the corporation. They face inward to interact with the firm's R&D and business operating units, in order to identify operating units' interests and priorities. CVCs support the corporation's existing businesses by introducing new technologies and partnerships to its operating groups. At the same time, CVCs help identify technologies and opportunities that fall between or beyond the corporation's existing businesses.

This report uses industry data and original survey data to describe trends and characteristics of CVC organizations and investments. These data provide insight on a range of issues relating to CVC operations and investments.

INTRODUCTION

Established corporations seeking growth today face an increasing need to innovate. While some corporations try to grow through external acquisitions, many recognize the ultimate importance of generating "organic growth" through innovation. Corporations are searching for new approaches to innovation.

Corporations use a variety of approaches to innovation, including internal R&D, incubation of new businesses, and strategic investments and alliances. Many are emphasizing a strategy of "open innovation" to bring external sources of innovation into the firm. Given the success of venture capital in creating new entrepreneurial companies and technologies, corporations have also looked to that model as yet another approach to innovation.

Corporate venture capital (CVC) may be defined as programs in established firms that make investments in entrepreneurial companies. Typically, a CVC makes a financial investment—just as independent venture capital does—and receives a minority equity stake in the entrepreneurial company. A CVC may also facilitate investment of in-kind and other resources into the portfolio company. In return, the corporation gains a window on both new technologies and strategically complementary companies that may become strategic partners.

While the sole objective of independent venture capital is financial return, CVCs generally have a strategic objective as well. That objective may include leveraging external sources of innovation, bringing new ideas and technologies into the company, or taking "real options" on technologies and business models (by investing in a wider array of technologies or business directions than the company can pursue itself).

Corporate venture capital may be viewed in the broader context of corporate venturing, including both internal and external venturing. Internal venturing programs "go inside" the firm and create entrepreneurial ventures from within the corporation. Entrepreneurial teams are granted independence and resources to innovate and form new business ventures. External venturing programs "go outside" the firm and tap external sources of innovation, whether through research collaborations with universities, strategic alliances with other firms, or partnerships with entrepreneurial companies. Often the firm's internal and external venturing efforts are closely related and interact with each other.

CVC programs in established corporations aim to build relationships with the entrepreneurial venture community. They learn about technology and business directions of strategic interest, and then make investments that create new strategic opportunities for the corporation. By interacting with the firm's R&D and business operating units, CVC programs identify the operating units' interests and priorities. CVCs support the corporation's existing businesses by introducing new technologies and partnerships to its operating groups. At the same time, CVCs help identify technologies and opportunities that fall between or beyond the corporation's existing businesses, which could then form the basis for new business directions.

The following chapters use industry data and original survey data to describe trends and characteristics of CVC organizations and investments. In 2007, the National Venture Capital Association (NVCA)—in partnership with researchers at the Wharton School and the MIT Sloan School—conducted a survey of 48 CVC organizations. These resulting data provide insight on a range of issues relating to CVC operations and investments.

1. Investment Trends (1995–2006)

CVCs have become a significant part of overall venture capital activity. In the year 2000, at the peak of the most recent venture capital cycle, more than $100 billion in venture capital was invested. About 16 percent of that investment was from CVCs. After 2002, total venture capital investment stabilized at around $20 to $25 billion per year, and CVC investment stabilized at around 6 to 8 percent of total venture capital investment.

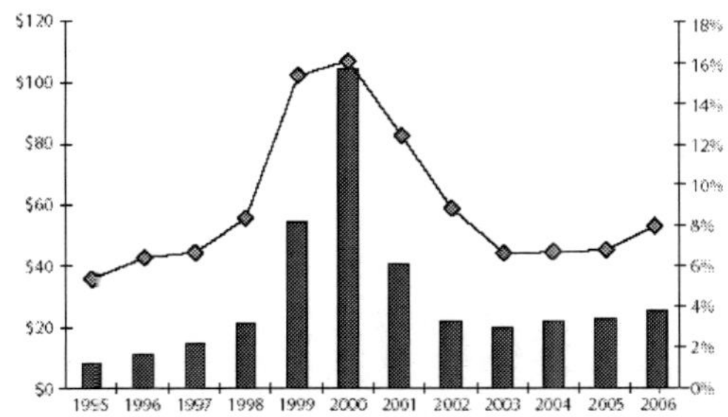

Bubble: Year 2000

- Total VC = more than $100 billion
- CVC share = about 16%

Post-Bubble: Years 2002–2006

- Total VC = about $20–$25 billion/year
- CVC share = about 6%–8%

Source: National Venture Capital Association

Figure 1-1. How Much CVC and Venture Capital Investment Is There?

Over the past several decades, waves of CVC activity have largely paralleled the cycles of overall venture capital investment activity. In recent years, CVCs have become a significant part of overall venture capital activity.

Figure 1-1 depicts overall venture capital investment and CVC investment from 1995 to 2006. In this period, the "internet bubble" of 1999–2001 stands out as the latest and most extreme example of boom-and-bust cycles that have characterized venture capital investing over the past several decades.

In the year 2000, the CVC share of total venture investment rose and peaked just as total investment did. This suggests that many CVCs acted as "latecomers," entering the market or increasing investment just as the "bubble" was about to burst. After 2002, total venture capital investment stabilized and resumed the longer-term trend line from 1995 to 1998. At the same time, CVC investment also stabilized as a percentage of total venture capital investment.

Investment Distribution by Company Stage of Development

Generally, the sole objective of independent venture capital is to generate financial returns to investors. By contrast, the investment objective for corporate venture capital—affiliated with a parent company that is typically a nonfinancial operating company—is a combination of financial returns and strategic benefits to the parent company.

The strategic benefit to the parent company may be to gain a window on new technology, take "real options" on new technologies or markets, or form a strategic alliance with a company with complementary technology or capabilities. Depending on the strategic objectives, CVCs may make investments in earlier-stage technologies and companies or later- stage technologies and companies.

Table 1-1 lists the definitions used by the National Venture Capital Association.

Figure 1-2 shows the distribution of venture capital and CVC investment in 2006 by the stage of development of the company receiving the investment. In 2006, as a share of investment, CVCs invested less in startup and early-stage companies, relative to all venture capital.

The share of venture capital and CVC investment in startup and early-stage companies has fallen in recent years. From 1998 (the last "normal" year before the 1999–2001 bubble) to 2006, the share of venture capital investment in startup companies fell from 8 percent to 5 percent, and the share of investment in early-stage companies fell from 26 percent to 15 percent. For CVCs, the share of investment in startup companies fell from 7 percent to 3 percent, and the share of investment in early-stage companies fell from 23 percent to 10 percent.

Table 1-1. Stages of Company Development

Stage	Definition
Seed/Startup Stage	The company has a concept or product under development, but is probably not fully operational. Usually, the company has been in existence for less than 18 months.
Early Stage	The company's product or service is in testing or pilot production. In some cases, the product may be commercially available. The company may or may not be generating revenues. Usually, the company has been in business for less than three years.
Expansion Stage	The company's product or service is in production and commercially available. The company demonstrates significant revenue growth, but may or may not be showing a profit. Usually, the company has been in business for more than three years.
Later Stage	The company's product or service is widely available. The company is generating ongoing revenue and is probably cash-flow positive. It is more likely to be profitable, but not necessarily.

Pre-Bubble: Year 1998 – VC and CVC Investment

Percentage of Investment		
Stage	**VC**	**CVC**
Startup	8%	7%
Early stage	26%	23%

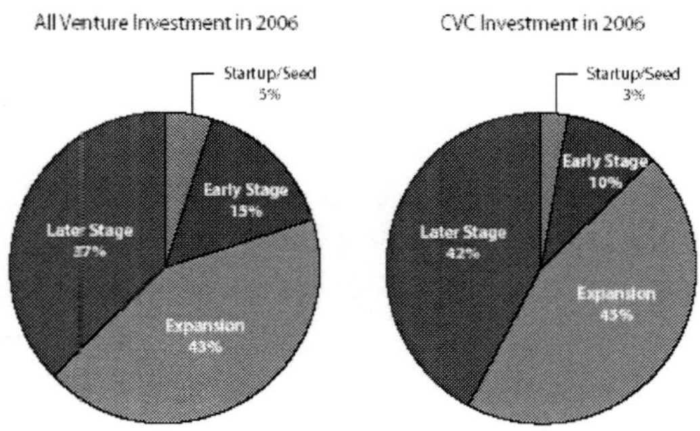

Post-Bubble: Year 2006 – VC and CVC Investment

	Percentage of Investment	
Stage	**VC**	**CVC**
Startup	5%	3%
Early stage	15%	10%

Source: National Venture Capital Association

Figure 1-2. How Does CVC Investment Compare with Venture Capital, by Stage of Development?

Investment Distribution by Industry Sector

Table 1-2 presents the industry sector distribution of venture capital and CVC investments in 2006.

Table 1-2. Which Industry Sectors Receive Venture Capital and CVC Investment?

Industry Sector	All Venture Investment 2006	CVC Investment 2006
Software	19.5%	13.4%
Biotechnology	17.6%	22.0%
Medical Devices and Equipment	10.4%	8.6%
Telecommunications	10.2%	12.0%
Semiconductors	7.8%	10.5%

Table 1-2. (Continued)

Industry Sector	All Venture Investment 2006	CVC Investment 2006
Industrial/Energy	6.9%	5.3%
Media and Entertainment	6.4%	10.1%
Networking and Equipment	4.2%	3.8%
IT Services	4.0%	3.5%
Electronics/Instrumentation	2.7%	4.8%
Business Products and Services	2.6%	2.1%
Consumer Products and Services	2.0%	0.8%
Financial Services	1.7%	1.2%
Computers and Peripherals	1.6%	1.5%
Healthcare Services	1.5%	0.3%
Retailing/Distribution	0.8%	0.2%
Other	0.1%	0.0%
Total	**100%**	**100%**

Source: National Venture Capital Association

Top Sectors for VC Investment

- Software
- Biotechnology
- Medical devices
- Telecommunications

Top Sectors for CVC Investment

- Biotechnology
- Software
- Telecommunications
- Semiconductors
-

In contrast, the top sectors for venture capital investment in 1998 were software and telecommunications. For CVC investment that year, the top sectors were software, telecommunications, networking, and media/entertainment. Between 1998 and 2006, the most significant changes in industry sector of investment, for both venture capital and CVC, were biotechnology and medical devices receiving increasing shares.

2. Strategic and Financial Objectives

While an independent venture capital fund's sole objective is making financial returns, a CVC typically has a combination of financial and strategic objectives. Generally, a CVC has a strategic mission to help "grow the business" of the parent company. It does this by helping the company identify new ideas or technologies, develop new products or processes, and enter new markets or enhance existing businesses.

Corporations seeking to innovate are emphasizing "open innovation" models that bring external ideas and technologies into the firm. As part of an open innovation approach, CVC programs interact with the venture capital community, and make investments in venture funds or companies. Through these interactions and investments, they learn about and gain access to technologies and ideas that reside in small entrepreneurial firms.

Many CVCs make investments as limited partner investors into independent venture capital funds. By doing so, a CVC develops experience in venture investing, builds relationships in the venture capital community, and gains a view on the "deal flow" of technologies and companies seeking venture capital funding. Most CVCs make direct investments and take minority equity stake in companies, just like independent venture capital funds.

Some CVCs may make investments exclusively for strategic benefits, without consideration of financial returns, For example, a CVC focused on learning about new technologies and bringing new ideas into the parent company may choose to invest without regard to financial returns. On the other hand, some CVCs may invest only for financial returns. Usually, a nonfinancial corporation would not make financial investments solely for financial return—in effect, entering into a financial investment business. But a corporation with a strong industry position may find that its unique knowledge of relevant markets and technology enables it to identify attractive opportunities to invest in smaller companies.

Most CVCs balance strategic objectives with the financial objective of making a financial return. For any individual investment, applying financial criteria in the decision imposes a financial discipline that helps ensure the investment's quality. Some would argue that a "strategic investment" in a company that fails financially is not likely very strategic. Looking at the CVC as a whole, experienced CVC practitioners recognize that ultimately the CVC has to "pay its own way." To keep the support of corporate management, the CVC not only has to deliver strategic benefits, but also cover costs and earn a reasonable return over time. In other words, a CVC that persistently loses money will not survive.

Figure 2-1 presents survey results on direct investments and limited partner investments by CVCs in 2004–2006. The results indicate that many CVCs make limited partner investments in venture capital funds, even as they make direct investments in companies.

Figure 2-2 illustrates survey results about the mix of strategic and financial objectives of CVC organizations. The most common CVC investment approach is to screen investment proposals for strategic value to the parent company and then, conditional on the proposal having strategic value, apply standard financial investment analysis similar to any independent venture capital fund that pursues financial returns.

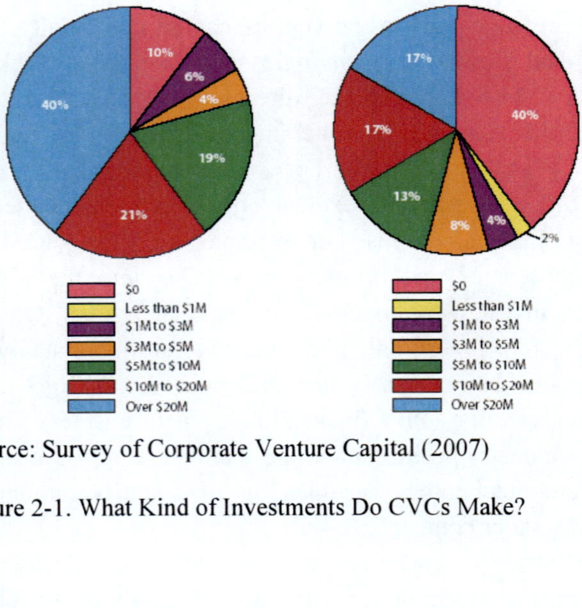

CVC Investments

- 90% of CVCs made direct investments
- 60% of CVCs made limited partner investments

Source: Survey of Corporate Venture Capital (2007)

Figure 2-1. What Kind of Investments Do CVCs Make?

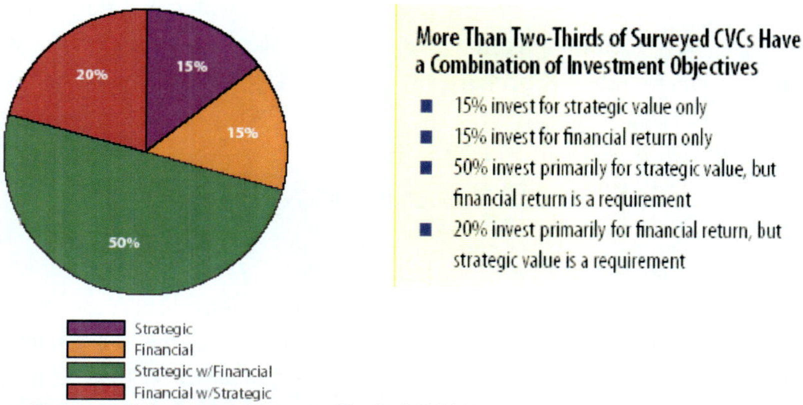

Source: Survey of Corporate Venture Capital (2007)

Figure 2-2. Do CVCs Invest for Strategic or Financial Objectives?

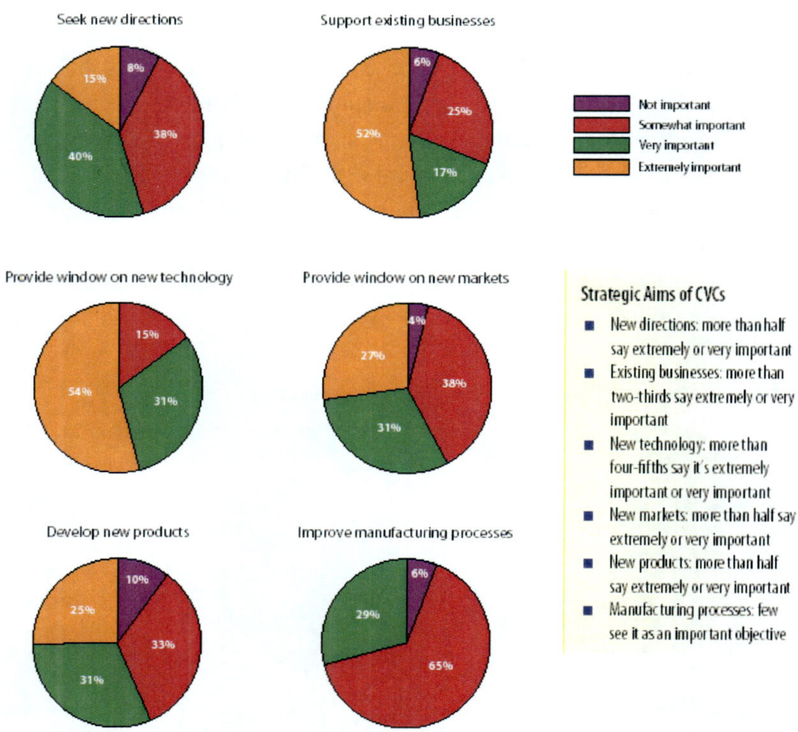

Source: Survey of Corporate Venture Capital (2007)

Figure 2-3. What Strategic Aims Are Important to CVCs?

Figure 2-3 presents survey evidence on the strategic aims of CVCs. Identifying new technologies that can advance the businesses of the parent company appears to be the primary strategic objective for most CVCs.

3. Structure and Organization

CVC programs are structured in various forms. Some CVCs are organized as independent subsidiary companies, while others operate as a group within the parent company organization. Some CVCs have a "dedicated" investment fund, where the corporation commits a given amount of capital for investment. Other CVC programs are "discretionary," in that investment capital is allocated as investment opportunities arise. Most CVC programs have a corporate-wide mission, receive corporate funding, and report to the corporate level.

Corporations have established CVC programs with a variety of structural and organizational forms. Some are independent subsidiaries, and some are groups within the parent company. Some have a "dedicated" investment fund structure, similar to independent venture capital, with a fixed amount of committed capital available for investment. Others are "evergreen" or "discretionary" in that investment capital is allocated as investments opportunities arise; there is no fixed fund structure similar to independent venture capital funds.

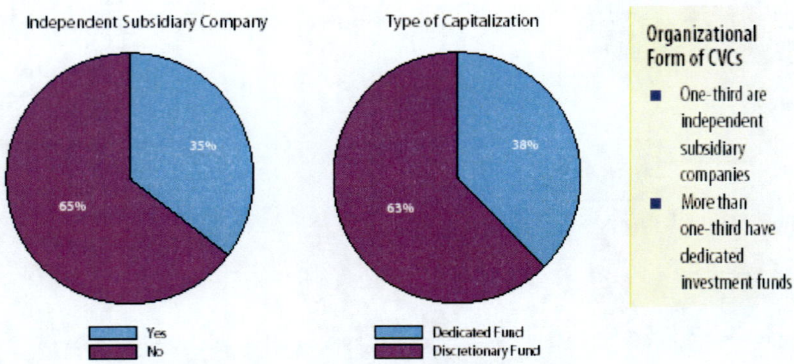

Source: Survey of Corporate Venture Capital (2007)

Figure 3-1. What Organizational Forms Do CVCs Take?

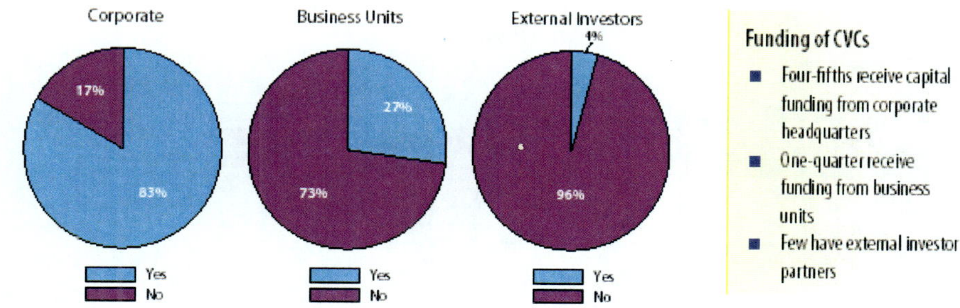

Source: Survey of Corporate Venture Capital (2007)

Figure 3-2. What Are the Sources of Capital for CVCs?

The source of capital for a CVC may be the corporate level of the parent company, business units within the parent company, or even possibly external investment partners. Most commonly, CVCs are funded by corporate headquarters. Often, business units may also contribute funding, especially when the CVC mission is closely aligned with a specific business unit. In some cases, a CVC may have limited partner investment relationships with one or more external independent venture capital funds.

Figure 3-1 depicts survey results about the organization and capital structure of CVCs in the sample.

Figure 3-2 shows the sources of capital for CVCs in the survey sample.

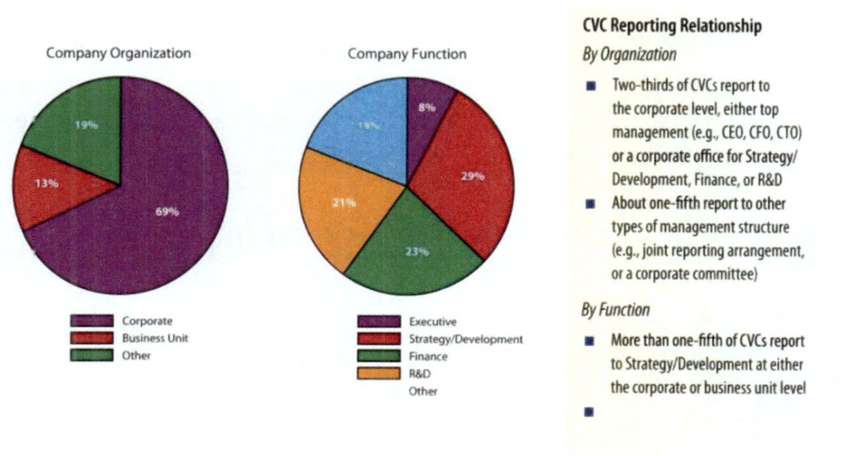

Source Survey of Corporate Venture Capital (2007)

Figure 3-3. Who Do CVCs Report To?

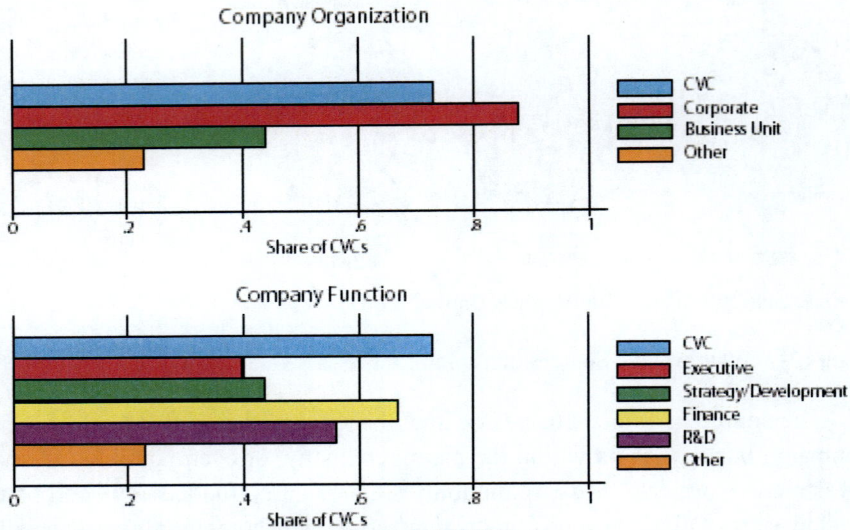

Source: Survey of Corporate Venture Capital (2007)

Figure 3-4. Who Is Represented on the Investment Board of CVCs?

Figure 3-3 shows different reporting relationships for CVCs in the survey. Most CVC programs have a corporate-wide mission and receive corporate funding, and report to the corporate level. CVCs in Japanese companies are most often associated with the R&D function. CVCs in pharmaceutical companies also often have strong ties to R&D.

Figure 3-4 shows the composition of investment boards for CVCs in the survey sample. The CVC investment board makes decisions on investments, and its composition reflects the balance of stakeholder and functional interests that enter into the decision making.

CVC Investment Board

By Organization

- Corporate-level offices and the CVC organization itself are most often represented on investment boards
- Business units are represented on the investment board in more than two-fifths of the CVCs

By Function

- Finance is represented on two-thirds of CVC investment boards
- R&D is represented on more than half of CVC investment boards

4. Investment Sourcing and Structure

CVCs and independent venture capital have developed working relationships that are mutually beneficial. CVCs communicate with independent venture capital to share information about areas of investment interest, and about companies and technologies seeking venture funding. CVCs benefit from access to the investment "deal flow" of independent venture capital, while independent venture capital benefits from strategic insight and technology expertise provided by CVCs. CVCs and independent venture capital also often co-invest in companies through syndicated investments.

Most CVCs make direct investments in companies, just as independent venture capital funds do. CVCs and independent venture capital have developed working relationships that are mutually beneficial, sharing information on investment opportunities and often co-investing together in deals.

Figure 4-1 shows results on the sourcing of CVC investment deals. The most important source of investment deals for the surveyed CVCs is independent venture capital, highlighting the relationship of CVCs with independent venture capital. Other significant sources of CVC investment deals include referrals from business units within the CVC's parent company and direct solicitation from companies seeking investment.

In the past, independent venture capital has often been skeptical of CVCs for perceived concerns that include: CVCs lack experience and sophistication in making investments; they are too slow in making investment decisions; they are unreliable and inconsistent in providing follow-on support to portfolio companies; and they have strategic interests that may conflict with a portfolio company's interests. In addressing those concerns, many CVCs now emphasize that they operate and act just as any independent venture capital fund, in that they invest for financial return and try to maximize the financial value of their investment. In effect, for a CVC, once an investment deal has passed a strategic "screen" or criterion for investment, it is treated just as a financial investment. As a financial investor and shareholder, a CVC is no different from an independent venture capital investor. Its objective is to maximize the portfolio company's value and act in the portfolio company's financial interest. CVCs are therefore careful not to encumber a portfolio company with restrictive agreements that benefit the parent company's strategic interest, but may turn out to be disadvantageous to the portfolio company.

In fact, CVCs and independent venture capital often co-invest in companies in syndicated investments. The independent venture capital investor usually takes the role of lead investor. That investor provides the financial and investment expertise for structuring and managing the deal, while the CVC joins the syndicate as a co-investor adding strategic value to the deal. In syndicated investments, independent venture capital investors can ensure that business agreements between the portfolio company and the CVC parent company are in the portfolio company's best interest and do not create a conflict of interest.

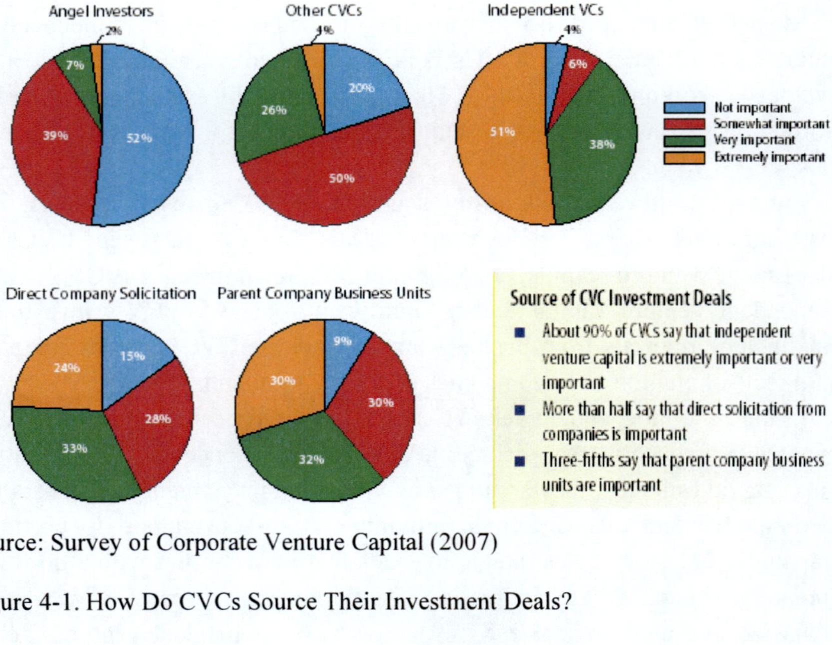

Source: Survey of Corporate Venture Capital (2007)

Figure 4-1. How Do CVCs Source Their Investment Deals?

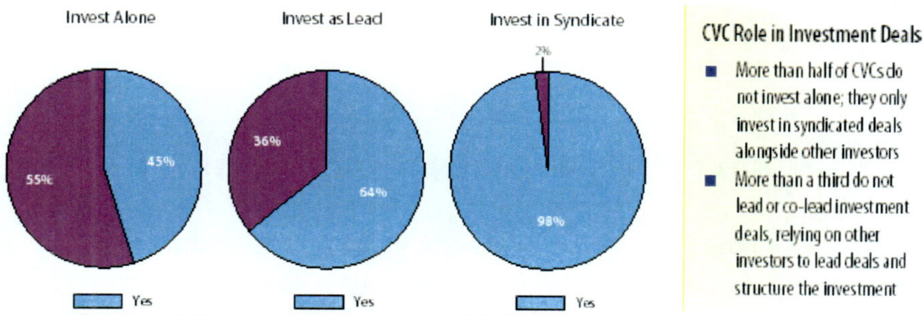

Source: Survey of Corporate Venture Capital (2007)

Figure 4-2. What Role Do CVCs Play in Investment Deals?

Figure 4-2 shows the structure of CVC investment deals and the role that CVCs play. Nearly all of the surveyed CVCs participate in syndicated investments, and more than half will invest only in syndicated deals. Many CVCs will not lead an investment deal, but rely on their investment partners to lead and structure the deal.

Upon making an investment, CVCs may take either a non-voting observer seat or a full voting member seat on the portfolio company's board. CVCs most often become non-voting observers since their primary interest is gaining a strategic view on technology and market developments at the portfolio company. CVCs with a stronger financial investment interest may be more likely to take a full voting member seat on the portfolio company board.

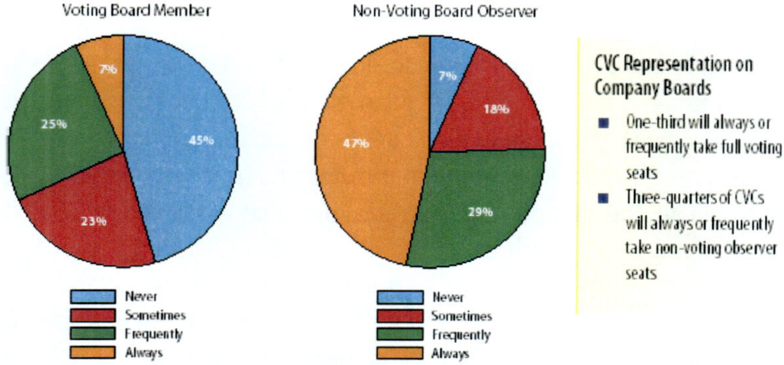

Source: Survey of Corporate Venture Capital (2007)

Figure 4-3. Do CVCs Take Seats on Portfolio Company Boards?

Figure 4-3 shows that most of the surveyed CVCs will take a non-voting observer seat on the portfolio company board, while many will take a full voting seat at least some of the time.

5. Interactions with R&D and Business Units

CVCs interact with the company's R&D and business units to understand the operating units' technology directions and business strategies. They then make investments that support or complement those efforts within the company. In screening investment proposals, CVCs draw on technical and business expertise from the operating units. An R&D or business unit may "sponsor" a CVC investment, indicating its support for the investment. C VC portfolio companies benefit from in-kind support and relationships with the parent company's operating units.

CVCs generally have a strategic mission to enhance innovation by bringing new ideas or technologies into the parent company. The specific approaches to the strategic mission vary widely, as do the specific forms of CVC structures and processes. Some CVCs are oriented to R&D organizations in the parent company, while others are oriented to the business operating units. Some CVCs are closely affiliated with a particular business within the parent company, while others serve all the businesses. Some CVCs focus on supporting the parent company's existing businesses, while others look to "white spaces" outside of existing business in order to identify new directions and opportunities.

But whatever the specific approach, in order to deliver strategic value to the parent company, a CVC needs to understand the technology directions and business strategies of the operating units in the parent company. In the investment process, the CVC communicates and interacts with R&D and business units in the company. Business units may provide "wish lists" of technology interests or priorities to a CVC, while a CVC may brief business units on new technologies from the entrepreneurial venture community. In screening investment proposals, CVCs draw on the technical and business expertise of the R&D organizations and business operating units. In many cases, an R&D or business unit may " sponsor" a CVC investment, indicating its support for the investment.

Figure 5-1 presents CVC survey responses about using technical or business personnel from the parent company to conduct due diligence on investment proposals. Most CVCs report that they use parent company personnel to a large extent or moderate extent.

In Figure 5-2, three-fifths of CVCs indicate that having an R&D or business unit in the parent company act as sponsor for an investment is **extremely important or very important to the CVC's** decision to invest.

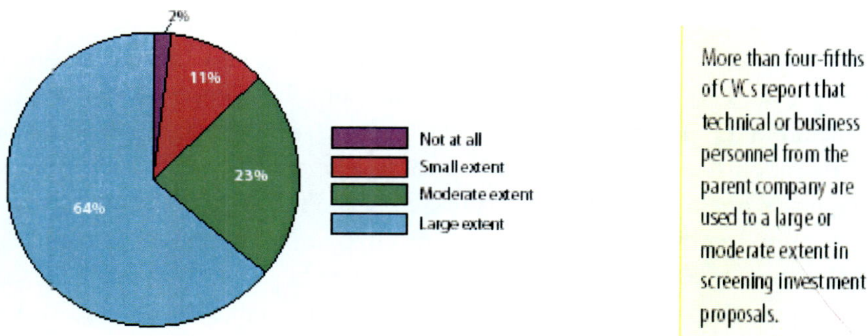

Source: Survey of Corporate Venture Capital (2007)

Figure 5-1. Do CVCs Use Parent Company Personnel for Due Diligence on Investment Proposals?

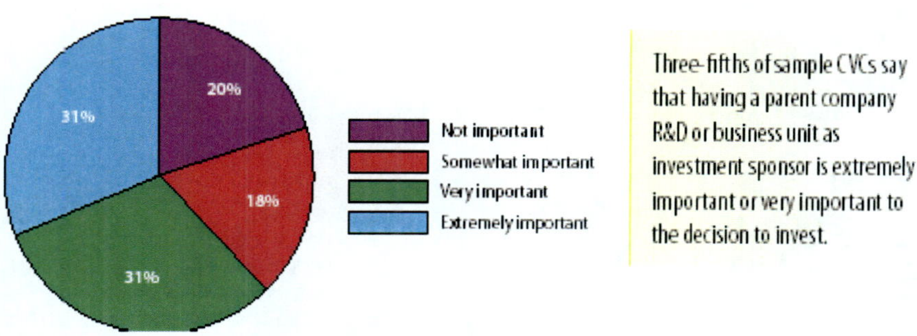

Source: Survey of Corporate Venture Capital (2007)

Figure 5-2. How Important Is It to Have an R&D or Business Unit Sponsor for a CVC Investment?

CVCs have to be careful to distinguish their role as an investor in a portfolio company from their role as an intermediary or conduit between the portfolio company and the CVC parent company. As an investor (and shareholder), a CVC seeks to maximize the value of the investment and acts in the portfolio company's interest. As an interface between the portfolio company and the CVC parent company, the CVC may provide information and introductions to help identify strategic interests and potential collaborations or alliances between the portfolio company and the CVC parent company. However, CVCs must maintain a clear separation, or "firewall," between the CVC and parent company in intellectual property, legal, or financial matters.

Where CVC investments are well chosen, there is a good strategic fit between the CVC parent company and the portfolio company, and strategic cooperation is likely to develop. In some cases, formal agreements for collaboration (such as joint R&D or co-development) with the CVC parent company's operating units are established at the time of CVC investment. In most cases, portfolio companies benefit from in-kind support and strategic relationships as a consequence of CVC investment.

Figure 5-3 shows that CVCs or their parent companies provide various forms of in-kind support to CVC portfolio companies.

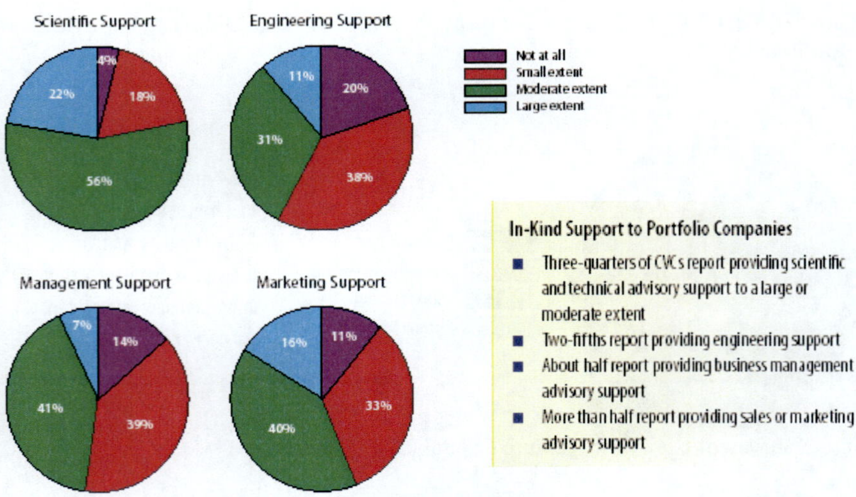

Source: Survey of Corporate Venture Capital (2007)

Figure 5-3. What Types of In-Kind Support Are Provided to Portfolio Companies?

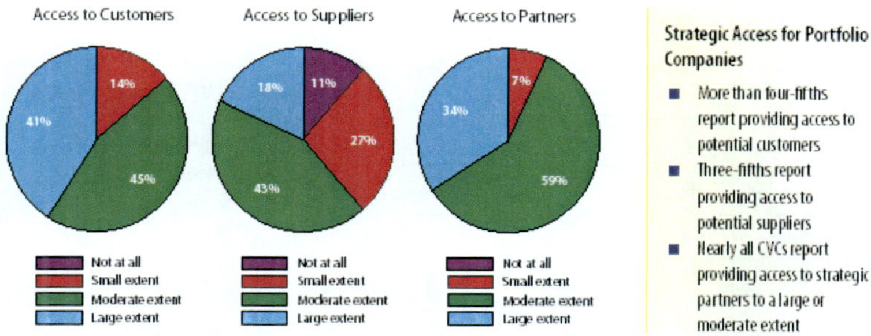

Source: Survey of Corporate Venture Capital (2007)

Figure 5-4. What Types of Strategic Access Are Provided to Portfolio Companies?

Figure 5-4 shows that CVCs or their parent companies help portfolio companies develop strategic relationships by providing access to potential customers, suppliers, and strategic partners.

6. Personnel and Incentives

A key issue for personnel management in CVCs is whether the organizational culture will be more similar to the parent company or to independent venture capital firms. For example, a CVC structured as an independent subsidiary company—with a dedicated investment fund and personnel hired from outside the parent company—will have an organizational culture that is more similar to an independent venture capital firm. Personnel compensation, especially the extent of incentive compensation, is a critical aspect of CVC personnel management.

Established corporations recognize that entrepreneurial companies are a dynamic source of innovation. They also recognize the success of venture capital as a model for financing and supporting entrepreneurial companies. To tap this external source of innovation, many corporations have adopted the venture capital approach and formed CVC programs to invest in entrepreneurial companies.

While independent venture capital seeks only financial return as its objective, corporate venture capital aims to fulfill strategic as well as financial objectives. As such, CVCs need to have staff that understand the technical and business strategy of the parent company, and also have knowledge and experience in venture capital investing.

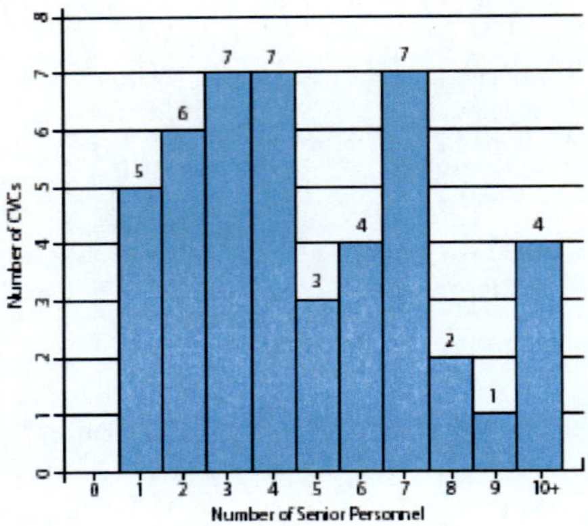

Number of CVC Senior Personnel

- Nearly a quarter of CVCs have only 1 or 2 senior personnel
- More than half have fewer than 5 senior personnel

Source: Survey of Corporate Venture Capital (2007)

Figure 6-1. How Many Senior Personnel Are There at CVCs?

Figure 6-1 shows the size of CVC organizations in the survey sample, in terms of the number of senior personnel. Most of the surveyed CVCs operate with a small staff of senior personnel.

Senior personnel at CVCs play a dual role: they reach outward to develop relationships with the venture community and also inward to communicate with business units about strategic interests and directions. Most CVC personnel have prior experience working in the parent company's operating units, so they "speak the language," know the people, and can communicate with the operating unit stakeholders.

To build partnerships between the CVC organization and operating units, CVCs may use personnel secondments. These are temporary developmental or rotational assignments (typically for one to two years) to bring people into the CVC organization from other parts of the company. CVCs are also likely to hire people from outside of the company. These hires bring expertise and capability in venture investing, along with relationships with the venture community.

Figure 6-2 shows the distribution of CVCs in the sample, by the percentage of senior personnel that are outside hires or on temporary assignment from other parts of the company. Nearly two-thirds of the CVCs have senior personnel hired from outside of the company, and more than one-quarter have senior personnel who are on temporary assignment from other parts of the company.

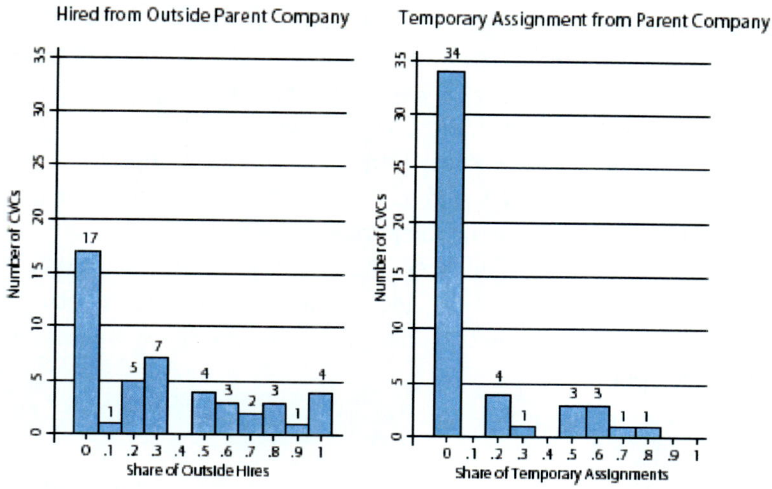

Source: Survey of Corporate Venture Capital (2007)

Figure 6-2. Where Do CVC Senior Personnel Come From?

Source of CVC Senior Personnel

- In nearly two-thirds of CVCs, 10% to 100% of senior personnel are outside hires
- In more than a quarter of CVCs, 20% to 80% of senior personnel are on temporary assignment from other parts of the company

The characteristic of personnel management that most influences the CVC's organizational structure may be personnel compensation. In independent venture capital firms, senior personnel receive "carried interest"—that is, a share of the investment returns—as a high-powered incentive in compensation. For CVCs within established corporations, such types of incentive compensation may be incompatible with the corporation's overall compensation system. Personnel at CVCs usually receive salary and bonus as compensation. Some CVCs provide a bonus structure tied to investment returns that approximates carried interest, and some directly provide carried interest.

Figure 6-3 shows the forms of compensation provided to senior personnel at CVCs in the survey sample. About one-third of CVCs offer compensation that either includes carried interest or a bonus similar to carried interest.

Personnel compensation is a key factor relating to personnel turnover. Successful CVCs with significant financial returns are likely to experience personnel turnover if compensation structures are inappropriate. Figure 6-4 presents evidence on the expected length of job tenure for senior personnel who are "permanent" (i.e., not on temporary assignment).

Overall, the evidence indicates that the surveyed CVCs successfully manage personnel compensation and organizational culture issues. CVC personnel find their organizational environment appealing and their compensation appropriate.

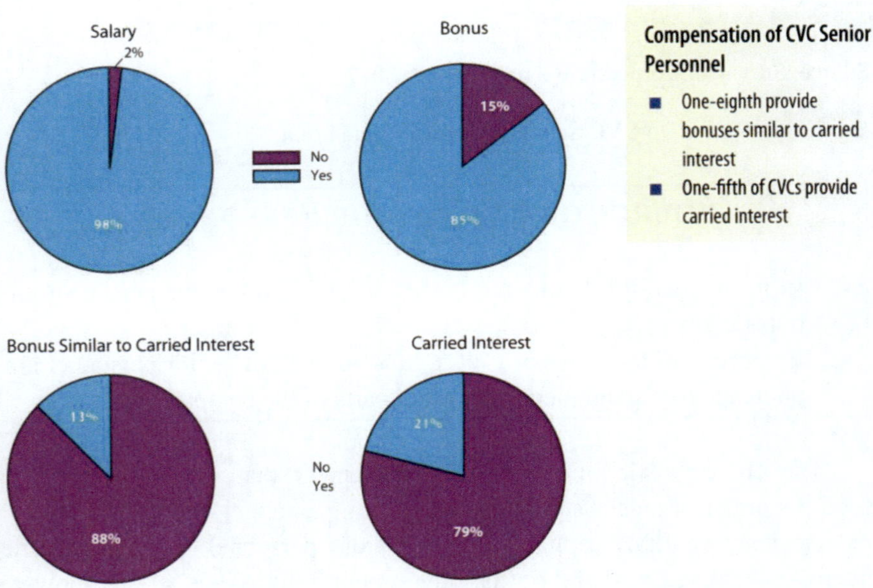

Source: Survey of Corporate Venture Capital (2007)

Figure 6-3. How Are Senior Personnel Compensated?

EXPECTED TENURE OF CVC SENIOR PERSONNEL

- In half of CVCs: 5 years or more
- In a third of CVCs: 3 to 5 years

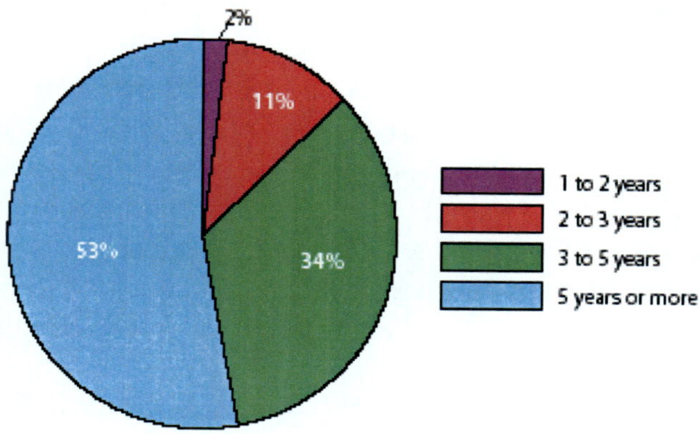

Source: Survey of Corporate Venture Capital (2007)

Figure 6-4. How Long Do Senior Personnel Stay at a CVC?

7. Investments and Strategic Value

CVC investments in venture-backed companies help the parent corporation learn about and access new technologies coming out of new firms. CVC investment in a company facilitates knowledge transfer and interaction between the company and the CVC parent corporation. Collaborations between a CVC portfolio company and operating units of the parent corporation may be established at the time of CVC investment or may develop over time as the portfolio company and parent company have informal contact and ongoing exchanges.

CVC programs in established corporations seek out entrepreneurial companies with new technologies that are of strategic interest. CVCs make investments to track technologies, and create "real options" on new technology or business directions for the corporation, that is, provide the corporation with opportunities to explore or pursue new directions.

CVCs usually work closely with R&D and business operating units in the company to identify technologies that are of interest to the operating units. CVCs aim to support the existing businesses of the company by introducing technologies or partnerships from venture-backed firms. At the same time, CVCs keep an eye on the "white spaces" that fall outside of the company's existing businesses, and look for technologies or opportunities that may become new strategic directions for the company.

CVC direct investments in companies help the parent corporation learn about and access new technologies coming out of new firms. CVC investments can facilitate knowledge transfer and interaction between a portfolio company and the CVC parent corporation. In some cases, collaborations between a CVC portfolio company and operating units within the parent corporation are established at the time of CVC investment. In most cases, collaborations form over time as informal contact and exchanges develop between the portfolio company and the parent company.

Figure 7-1 shows the number of direct investments made by CVCs during 2004–2006.

Figure 7-2 shows the number of collaborations formed between CVC parent companies and portfolio companies as a result of investments made during the 2004–2006 period. A striking result is that nearly all CVCs that report direct investments in 2004–2006 also report collaborations formed from investments in the same period.

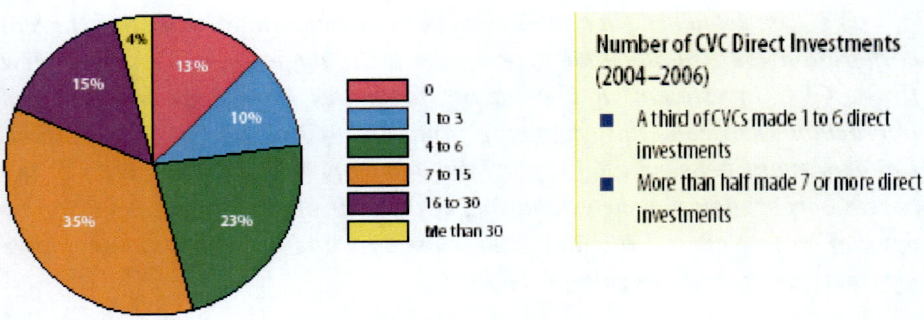

Source: Survey of Corporate Venture Capital (2007)

Figure 7-1. How Many CVC Direct Investments Were Made in 2004–2006?

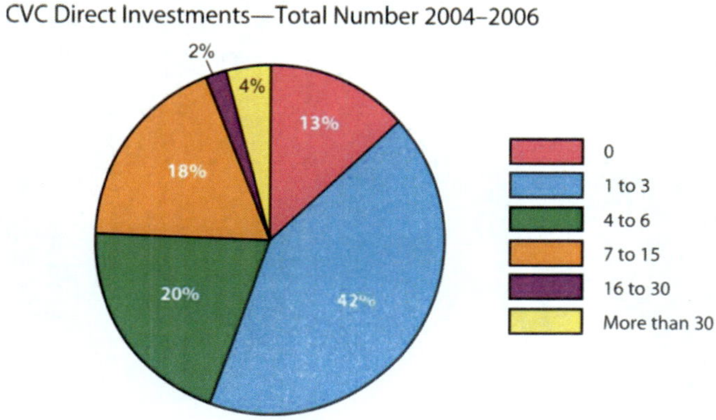

Source Survey of Corporate Venture Capital (2007)

Figure 7-2. How Many Collaborations Developed from CVC Investments in 2004–2006?

Figure 7-3 shows the types of collaborations that developed from CVC investments in 2004–2006. Among the CVCs surveyed, the most common type of collaboration formed between a CVC portfolio company and the parent corporation is an R&D collaboration.

The survey evidence indicates that CVCs bring strategic value to the parent corporation most often through R&D collaborations with CVC portfolio companies, and also often through commercial collaborations for licensing or sales.

NUMBER OF COLLABORATIONS FROM CVC INVESTMENTS (2004–2006)

- About two-fifths of CVCs report 1 to 3 collaborations
- More than two-fifths report 4 or more collaborations

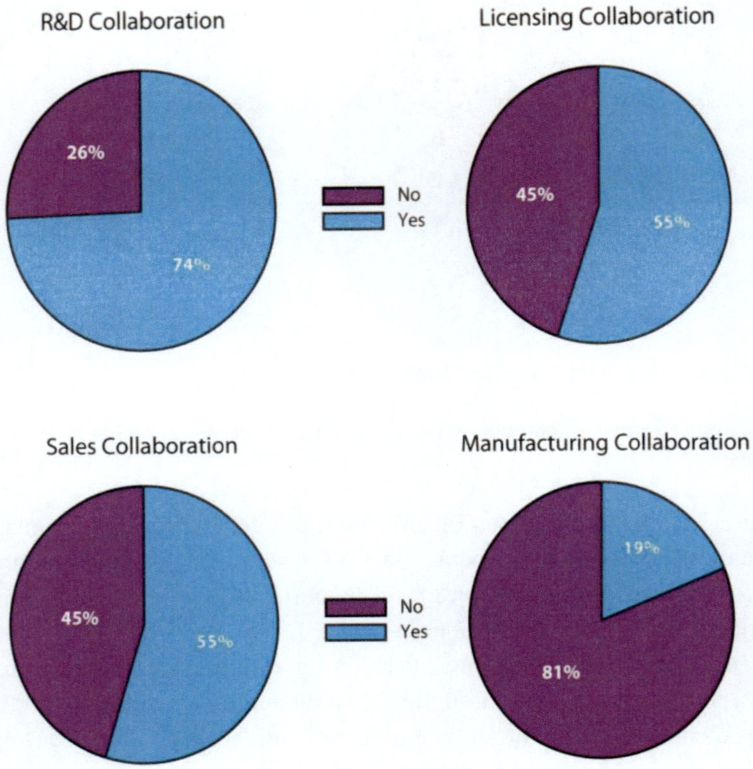

Source: Survey of Corporate Venture Capital (2007)

Figure 7-3. What Types of Collaboration Developed from CVC Investments in 2004–2006?

Type of Collaborations from CVC Investments (2004–2006)

- About three-quarters of CVCs report R&D collaboration
- More than half report licensing collaboration
- More than half report sales or distribution collaboration
- About one-fifth report manufacturing or other collaborations

8. Corporate Venturing

CVCs may be embedded in broader corporate venturing efforts and may interact with internal venturing programs. Corporate venturing includes both internal venturing programs and external corporate venture capital. Internal venturing programs give entrepreneurial employees resources and opportunities to create new ventures (e.g., seed funding for R&D or incubation resources to create new businesses). Successful internal ventures may eventually be incorporated into an existing business of the corporation, set up as a new business unit, or spun out as an independent company.

Corporations seeking innovation pursue a variety of "corporate venturing" activities that go beyond or outside the traditional approaches to R&D and business development. Corporate venture capital may be viewed in this broader context of corporate venturing, which includes internal venturing efforts as well as external corporate venture capital. While external CVC programs aim to identify and access external sources of entrepreneurial innovation, internal venturing programs seek to foster entrepreneurship and entrepreneurial innovation from within the company.

Internal venturing programs provide entrepreneurial employees with resources and opportunity to create new ventures: for example, seed funding for R&D, or incubation resources to create a new business. New internal ventures that become successful may eventually be incorporated into an existing business of the corporation, or established as a new business unit, or spunout as an independent company.

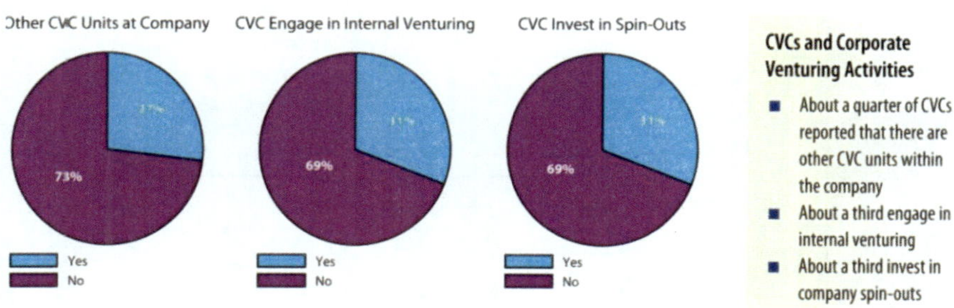

Source: Survey of Corporate Venture Capital (2007)

Figure 8-1. How Do CVCs Interact with Other Corporate Venturing Activities?

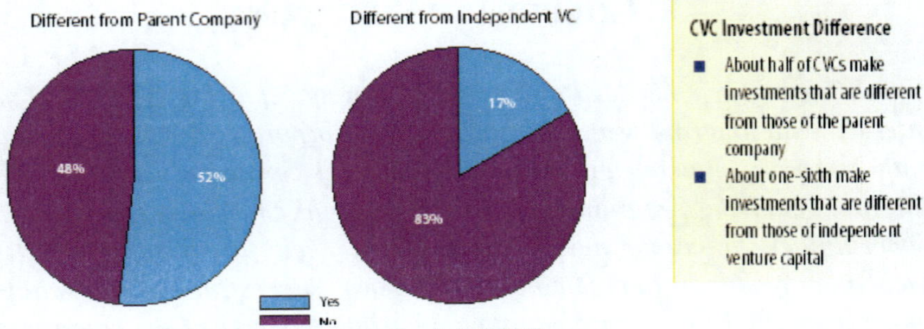

Source: Survey of Corporate Venture Capital (2007)

Figure 8-2. Are CVC Investments Different?

Corporations may take a variety of different approaches to internal and external venturing all at the same time. Multiple CVCs may exist within the same firm, for example, to focus on different technology areas or serve different business unit interests. CVCs may be embedded in broader corporate venturing efforts and interact with other internal venture programs.

Figure 8-1 indicates the corporate venturing context of CVCs in the survey sample.

Figure 8-2 shows survey evidence on whether CVCs make investments that are different from what a parent company or independent venture capital would usually make.

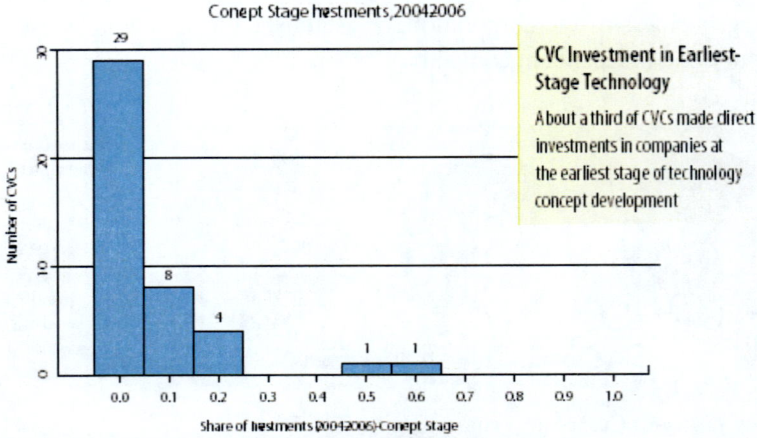

Source: Survey of Corporate Venture Capital (2007)

Figure 8-3. Do CVCs Invest in Companies at the Earliest Stage of Technology Concept Development?

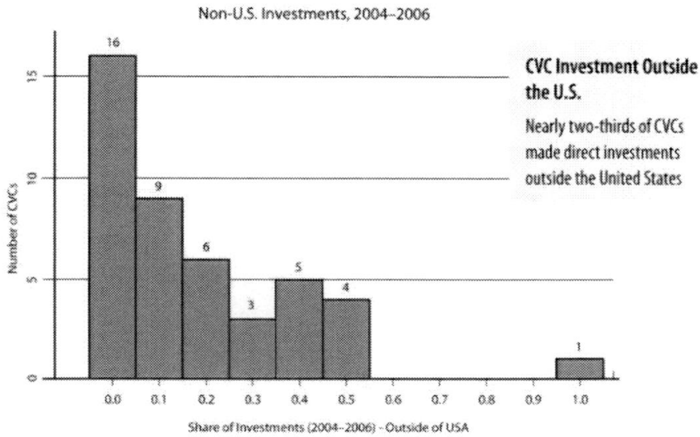

Source: Survey of Corporate Venture Capital (2007)

Figure 8-4. Do CVCs Invest in Non-U.S. Companies?

Half of the surveyed CVCs report that they do make investments different from their parent company. They may invest, for example, in earlier-stage technologies or technologies outside areas typical for the parent company. On the other hand, fewer CVCs report that they invest differently from independent venture capital. This may be because CVCs operate similarly to independent venture capital and often co-invest along with independent venture capital in syndicated investments.

Figure 8-3 shows how many of the surveyed CVCs made direct investments during 2004–2006 in companies at the earliest concept or intellectual property stage of technology development.

Figure 8-4 shows that nearly two-thirds of the surveyed CVCs made direct investments outside the United States. These survey data generally illustrate how CVCs fulfill their corporate venturing role by seeking technologies that are further afield and longer term.

9. OUTCOMES AND METRICS

There is no single "one-size-fits-all" approach to CVC that applies to all corporations. CVC best practices and performance assessment will depend on the CVC's role in the corporation, its mission, and the corporate environment. Assessing the CVC's strategic value requires judgments on

ultimate outcomes, such as the significance of insights the CVC provides to top management and operating units of the corporation; the importance of referrals, introductions, and external contacts and relationships that the CVC facilitates; and the strategic impact on the corporation of collaborations and partnerships that develop from CVC investments.

Unlike independent venture capital, which has a single objective of maximizing financial return on investment, most CVCs have a combination of strategic and financial objectives. For CVCs, assessing financial performance of investments is relatively straightforward in principle, but in practice, it may be problematic since many CVCs do not have a long enough track record of investing to assess financial returns.

Assessing the strategic value the CVC contributes to the wider corporation is perhaps more difficult. CVCs aim to form relationships with the external venture community, identify new technology and business directions that may be of strategic interest to the corporation, and create strategic options for the corporation by making investments in entrepreneurial companies. A CVC's strategic value could be measured in a number of ways, including: the quantity and quality of information provided both to top management and to the corporation's operating units; the significance of referrals, introductions, external contacts, and relationships facilitated by the CVC; and the strategic impact of collaborations and partnerships that develop from CVC investments in portfolio companies.

While some aspects of CVC performance may be directly measured (e.g., the number of investments made), a CVC's strategic value depends in large part on the subjective judgment of CVC "stakeholders." Figure 9-1 presents survey data on the extent to which management stakeholders within the parent corporation perceive the CVC as contributing value to the company.

PERCEPTION OF CVCS' VALUE CONTRIBUTION TO THE FIRM

- Fourth-fifths of CVCs report that senior management perceives them as contributing value to a moderate or large extent
- More than two-thirds report that technical management and business management at operating units perceive them as contributing value

Source: Survey of Corporate Venture Capital (2007)

Figure 9-1. Which Key Stakeholders Perceive That CVCs Contribute Value?

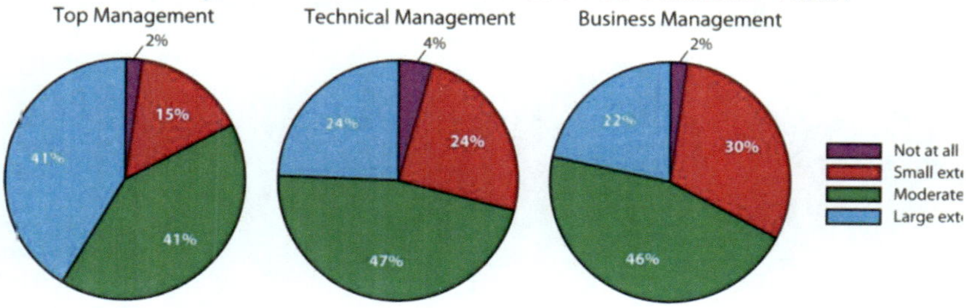

While communications regularly take place between investors and ventures, the number of face-to-face meetings relates to perceived VC value (Sapienza, 1992). Venture capital providers use funding rounds as a means of both managing their risk and providing knowledge and advice to their PCs. Such information sharing can include both explicit and tacit knowledge. Explicit knowledge may include identification of opportunities, names of client prospects, and information on potential alliance partners. Not every member of the TMT can obtain tacit knowledge directly; groups develop a common understanding via face-to-face interactions (Athanassiou & Nigh, 2000). Contact and interaction in financing rounds facilitates the transfer of tacit market knowledge to the TMT of new ventures. More rounds of financing provide more contact and communication, and thus provide opportunity for information exchange of this type. Such information can aid the new venture in identifying and developing foreign opportunities and increasing foreign sales.

Not all ventures that obtain CVC receive the same number of rounds that include corporate funding. Ventures receiving more rounds of CVC are more likely to have a high degree of internationalization.

> *Hypothesis 5: Conditional on receiving CVC funds, new ventures receiving more rounds of corporate funding are more likely to have a high level of international intensity than those receiving fewer such funding rounds.*

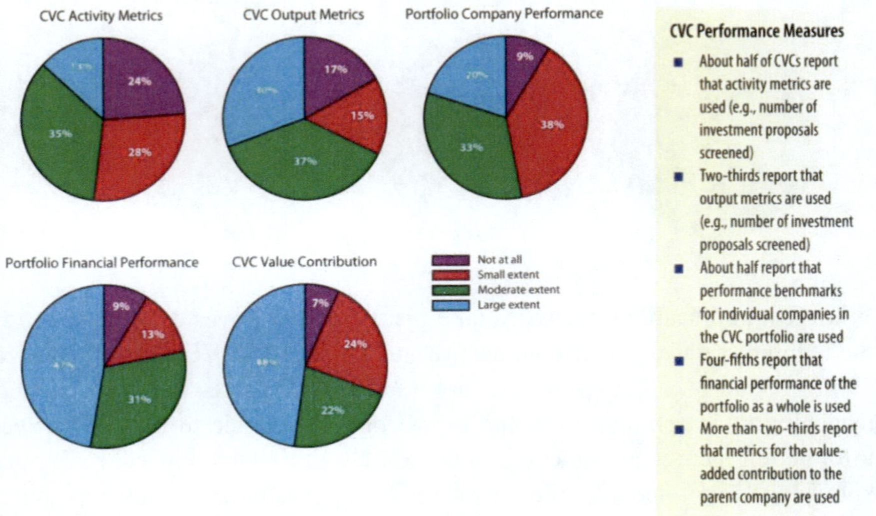

Source: Survey of Corporate Venture Capital (2007)

Figure 9-2. How Is CVC Performance Measured?

Figure 9-2 presents survey data on the types of metrics used by top management to assess CVC performance.

Activity metrics and output metrics are most easily measured, but relate only indirectly to ultimate outcomes for the parent company in terms of financial or strategic impact. Company-specific benchmarks and overall portfolio financial performance are metrics that an independent venture capital fund would use to assess and manage its investment portfolio.

The CVC's value-added contribution to the organization may include technology insights and briefings presented to R&D and business units, referrals and introductions to external contacts to assist R&D and business units, and collaborations and partnerships formed as a result of CVC investments.

Over the years, many CVCs have found that maintaining top management support for the CVC is difficult. During upswings in the venture capital investment cycle, many corporations are drawn into forming CVCs, but then quickly abandon the effort when the venture capital cycle turns downward. Some corporations, however, have shown a sustained commitment to corporate venture capital over the long term, maintaining funding support through the ups and downs of the venture capital cycle.

Figure 9-3 indicates the degree of funding support in 2004–2006, following the collapse of the "internet bubble," and the anticipated funding support for future investment in 2007–2009.

CVCs reporting stronger positive assessment of CVC value by top management (see Figure 9-1) are more likely to report increased support in the past and anticipate more investment in the future (see Figure 9-3). Top management perception of CVC value appears to translate into funding stability and support.

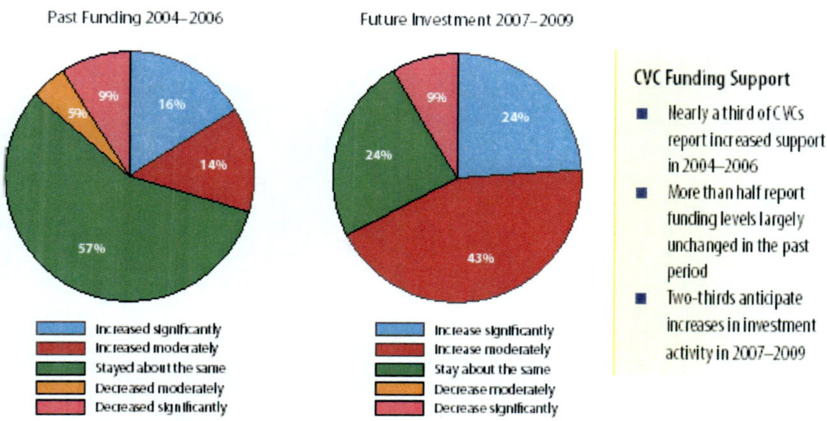

Source: Survey of Corporate Venture Capital (2007)

Figure 9-3. How Much Funding Support Do CVCs Have?

INDEX

A

B

C

I

J

K

L

M